A
D

I
N
F
I
N
I
T
U
M

.

.

.

The

Ghost

in

Turing's

Machine

A

D

I

N

F

I

N

I

T

U

M

.

.

.

S T A N F O R D   U N I V E R S I T Y   P R E S S

STANFORD, CALIFORNIA

Taking

God

out of

Mathematics

and Putting

the Body

back in

. . . .

An essay

in corporeal

semiotics

by

Brian Rotman

Stanford University Press

Stanford, California

© 1993 by the Board of Trustees of

the Leland Stanford Junior University

Printed in the United States of America

CIP data appear at the end of the book

Original printing 1993

Last figure below indicates year of this printing:

03   02   01   00   99   98   97   96   95   94

Stanford University Press publications

are distributed exclusively by Stanford University

Press within the United States, Canada, and

Mexico; they are distributed exclusively by

Cambridge University Press

throughout the rest of the world.

To

my father,

old Joe,

whose insistence

on speedy reckoning

& mental arithmetic

exposed me, early

in my childhood,

to the rigors

& wonder of

number

Across the span of Western thought, infinity has been a notoriously troublesome idea, difficult to pin down, full of paradox, and seemingly connected in some way or other with the divine. But whatever its philosophico-theological obscurities and contradictions, infinity in *mathematics*, as a phenomenon and an effect, is neither difficult to pin down nor hard to come by. One meets it immediately in elementary situations when, for example, one tries to divide a number by zero or compute the tangent of 90 degrees or express the fraction 1/3 as a decimal, or fundamentally, when one writes the ideogram ". . ." of mathematical continuation to signal that the progression of whole numbers 1, 2, 3, . . . be continued without end. And one meets it just as immediately in non-elementary mathematics. It is at work in the very idea of the geometrical continuum of points on a line and their integer-based real number descriptions—two linked abstractions which ground all post-Renaissance mathematics. And it is the founding signified, the crucial ontological term, in contemporary mathematics' description of itself as an infinite hierarchy of infinite sets.

How, one might ask, has mathematics so successfully tamed and incorporated the infinite? Incorporated it moreover at such a basic, elemental level, in so all-pervasive a way? How does infinity get to be an exact, rigorously specified mathematical object—an object about which mathematics delivers "true" and "objective" knowledge? Mathematics starts from the integers. Its entire formalism opens out from the sequence 1, 2, 3, . . . that mathematicians call the "natural" numbers. The question can therefore be particularized: What does it mean to say of these numbers that they are infinite, that they form a progression which is *endless*? In what sense are they *natural*, that is to say, before, independent, and outside of us? Precisely this question frames the present essay.

Of course, part of any answer is historical. The present-day interpretation of the mathematical infinite has a long story behind it; from Anaximander and Zeno, through Euclid, Aristotle, Carl Friedrich Gauss, culminating in Gottlob Frege's logic and Georg Cantor's infinities articulated within an axiomatics that dominates the twentieth-century mathematical scene. A century ago, as part of the initiating phase of this axiomatics, Richard Dedekind asked how numbers were to be defined. And he did so by means of a double question—"Was sind und was sollen die Zahlen?"—what are [*sind*] the numbers and what might/should [*sollen*] they be? Dedekind's answer—that numbers were a certain sort of sets—is now part of the very picture of a "naturalized" infinity of numbers that will be put into question here. Nevertheless, we might adapt his question, replacing his modality of being by a modality of being *thought*, and ask, "What is the infinite and how are we to think it?"

But, as we shall see, *being thought* in mathematics always comes woven into and inseparable from *being written*. We are never presented with the pure idea of infinity as such. How could we be? Instead, we meet only with certain mathematical inscriptions that in turn are connected to other inscriptions through a complex and open-ended spread of sense and interpretation. Thinking in mathematics is always through, by means of, in relation to the manipulation of inscriptions. Mathematics is at the same time a play of imagination and a discourse of written symbols. Should we not then pose the question of the mathematical infinite as a question of language, as part of an overall study of the nature and practice of mathematical signs—as part, that is, of a semiotics of mathematics? The answer is that we should, but that—obviously enough—we need to develop a semiotics of mathematics in order to do so.

In this sense this book starts where a previous one on mathematical signs—*Signifying Nothing*—felt itself able to sign off. In that text, having chased zero and zero-like signs across different cultural codes,

I called a halt at its presence within the binary formalism of Boolean logic governing the contemporary computer:

> To pursue zero further would be to have to say more about its role at the origin of this formalism. But such a project would require a critique of mathematical logic. In particular, one would need to unravel the assumptions behind the claim that is made for Boolean logic, with its referential apparatus of truth and falsity, to be *the* grammar of all mathematical, hence all scientific-technical, hence all supposedly neutral, culturally invariant, objective, true/false assertions about some prior "real" world. To do this would require a semiotics that went beyond zero to the whole field of mathematical discourse. A semiotics which, in order to begin at all, would have to demolish the widely held metaphysical belief that mathematical signs point to, refer to, or invoke some world, some supposedly objective eternal domain, other than that of their own human—timebound, changeable, subjective and finite—making. (1987: 107)

What I present here is not a further stalking of zero but a pursuit of infinity. However, the intimacy—both historical and conceptual—of the coupling between the two, between nothing/zero and all/infinity, manifest for example in the immediate production of infinity as the reciprocal of zero, makes any explication of the mathematical infinite a supplement to that of zero. Neither is what follows a philosophical critique of the metaphysical system, the rampant Platonism, that threads its way through the contemporary interpretation of mathematics—though, naturally, I cannot avoid engaging with Platonism, given its near universal acceptance. Rather it is a critique, in the form of an interrogation from first principles, of what seems to me an altogether more subtle metaphysical principle that permeates the entire subject. This is the principle of *ad infinitum* continuation that is inseparable from the mathematical community's wholesale acceptance of the view that the numbers are "natural," and its failure to ask the question of where these numbers could possibly come from. With this failure comes its inability to perceive either the need or the point of asking what the infinite is and how we are to think it.

In what follows I offer a semiotic model of mathematics of the kind required, namely one that extends to the "whole field of mathematical discourse." I read mathematical signs in terms of a certain written practice, a business of manipulating inscriptions that characterizes mathematical thought as a kind of waking dream. And, it turns out, one doesn't have to attempt a full-scale demolition of Platonism (which in any case could never succeed) in order to begin at all. Rather what happens is the reverse: it is the semiotic model—constructed, to be sure, despite and against the grain of that belief system—that enables one to shuck Platonism off in the end as a theological obfuscation of "number."

Finally, yes: mathematical signs, materially present to the senses of an embodied sign-user, are indeed of "human—*finite*—making." But what is this to mean? Isn't everything—everything corporeal—finite? Indeed it is; but pursuing the question of writing/thinking of the infinite only raises the question of "finite" and how we are to think "human making" in a new way. In the case of numbers, this way points beyond the comforting fixity of the finite-as-being within the infinitude of God—the view that the whole numbers in all their endlessness are and were always there before us, natural and God-given objects—to a view of numbers as open-ended and unfinished, as becoming, needing always to be counted into an actualizable form of "being."

Thanks to Memphis State University for the use of their library facilities; to the many audience members at various universities from Brown to Berkeley across the United States, whose questions and objections helped to improve the presentation and clarify the ideas here; to Barbara Herrnstein Smith for her acute and useful comments on an earlier draft; to Helen Tartar and the staff at Stanford University Press for their interest and enthusiasm in the project; to the American Council for Learned Societies for a research fellowship which enabled me to live in modest comfort whilst working on the final draft; and to Stanford University Humanities Center, whose award of a fellowship allowed completion of the work in the wonderfully congenial atmosphere of the Center.

A
C
K
N
O
W
L
E
D
G
M
E
N
T
S

CONTENTS

Ad

Infinitum

· · ·

The

Ghost

in

Turing's

Machine

God,

Number,

the Body

century ago the mathematician Leopold Kronecker declared that "God made the integers, the rest is the work of Man." This much cited remark was intended as a call for constructivist rigor, a polemic against the contemporary embrace of infinitism, a Pythagorean reaffirmation of the originating and privileged status of the whole numbers. And so it has been interpreted. It was not, one feels, meant with any theological literalness. And yet . . . is there not in the very idea of their *endlessness*, their continuation *ad infinitum*, something strange and other about the whole numbers, the imprint or trace of some disembodied, transcendental maker, perhaps?

Potential/Actual

The history of mathematics and the history of "the endless" intertwine. Issues of the infinite, of the nonfinishing, that which cannot be limited, bound, or traversed, arise as soon as mathematics takes quantity and length as its constituting abstractions and is obliged to make self-reflective sense of the questions "how many numbers?" and "how long, how divisible a line?"

There have been various moments of conflict and dislocation in the history of mathematics when the question of the infinite—what it means and how one is to think it—has pushed itself to the foreground of mathematical discourse. Each such moment arose in the wake of a speculative mobilization of infinitary reasoning which had resulted in obscurity and absurdity intolerable to mathematical thought.

The earliest crisis, associated with Zeno's infamous paradoxes, centered on whether one could assume space and time were *endlessly* divisible or whether, on the contrary, they were composed of ultimate and indivisible quanta. Zeno, attempting to establish the unreality of motion and plurality in the name of Parmenides' changeless and indivisible One, forced each of the possible alternatives into a contradiction. The effect of his arguments on classical thought was twofold: an avoidance of any appeal to motion or to the "endless" within their reasoning on the part of Greek mathematicians; and Aristotle's distinction, arising out of his engagement with Zeno's paralogisms and fundamental in all subsequent discussions, between a safe and legitimate *potential* infinite, an endless coming into being, and a dangerous, paradox-infested completed or *actual* infinite.

During the sixteenth and seventeenth centuries Aristotle's interdiction of actual infinity was set aside in the development of the theory of infinite series and the infinitesimal calculus. The second major problematic of the mathematical infinite then arose out of inconsistencies and absurdities, notably about infinitesimal magnitudes, that emerged within these theories. The attempt to eliminate these produced a movement toward rigor at the beginning of the nineteenth century associated with Carl Friedrich Gauss and Augustin Cauchy which repudiated the use or mention of an actual or completed infinity in any acceptable mathematical context. In less than a century the repudiation (despite Kronecker's Pythagorean campaign for constructivist renewal) was swept aside with the acceptance of Richard Dedekind's description of the continuum and Georg Cantor's theorization of the actual infinite—the set of all integers, all subsets of integers, all fractions, all points on a line, and so on—as bona fide mathematical objects.

The third crisis, whose "resolution" forms the horizon of present-day thinking about the mathematical infinite, emerged from the upheaval in the logic and ontology of number and the paradoxes that occurred in relation to Cantor's theory of infinite sets. Within this horizon two opposed accounts of mathematics emerged: the dominating orthodoxy of a Platonism inspired by Gottlob Frege and Jan Brouwer's anti-Platonic constructivism.

For most mathematicians (and, one can add, most scientists)

mathematics is a Platonic science, the study of timeless entities, *pure forms* that are somehow or other simply "out there," preexistent objects independent of human volition or of any conceivable human activity; mathematicians discover but never in any sense *invent* their presence and their properties. Accordingly the integers exist—out there—in their entirety as a single set, an unproblematic completed infinity; likewise the set of all subsets of the integers, the set of all subsets of these, and so on, ad infinitum. Mathematics is thus seen to be organized as an infinite hierarchy of infinite sets within a rigorously axiomatic framework by means of which all doubts, circularities, and antinomies have been swept away.

Against this there is the constructivist dissent. Brouwer disputed the coherence of classical, Platonic logic—rejecting the law of excluded middle that dictates that for any mathematical object, O, either O exists or O doesn't exist—by insisting that any mathematical proof of the existence of an object had to be in the form of instructions for arriving at it—that is, a finitely specifiable procedure that could "in principle" be executed in the mind. The progression of integers for Brouwer was a potential and not an actual infinity, available in principle as an endlessly producible sequence of purely mental acts, constructions to be performed deep inside our—Kantian—intuition of time.

The Natural Numbers

Now, as far as the purely logical engendering of the hierarchy of infinities is concerned, Kronecker's Pythagorean point is well put: evidently all infinities, potential as well as actual, whatever differences in psychology, logical coherence, meaning, and ontology constructivism and Platonism might attach to them, are rooted in the integers, the progression of objects that mathematicians, Platonists, and constructivists alike, call the *natural* numbers. "Natural" because they are given at the outset, taken for granted as a founding, unanalyzable intuition, outside any critique that might demand an account of how they come or came—potentially or actually—to "be." Indeed, for Platonism the very call for such critique would be senseless, whilst for

constructivism the issue is essentially dodged—sold short by constructivism's immersion in an unexaminedly ideal mentalism.

Evidently, if we are to understand the numbers' relation to God, it
is precisely their "endlessness," as natural and given or as the result of
an ideal construction, that needs to be interrogated. But faced with
Platonist orthodoxy's inability to question the integers' engenderment
and the complicity of its constructivist critics, how are we to escape
from the sedimented legitimacy and beguiling immediacy of the "natural"? How refuse the claim that numbers are elemental, mathematical ur-constituents of the actual or potential order of things? How
deny that producing or perceiving the endless progression of them is
inextricable from the apparatus of rational thought itself? Where *do*
numbers come from? If not from Kant's transcendental intuition or
Kronecker's God, then where?

What seems universally accepted is that numbers are inconceivable—practically, experientially, conceptually, semiotically, historically—in the absence of *counting*. Counting, whether with fingers,
pebbles, notches, tally marks, abacus beads, or notations on chalk-
boards, paper, and computer screens, is an activity involving *signs*.
And, as an activity, counting works through—it *is*—significant repetition. How are we—through what discursive apparatus and technology
of symbolic persuasion—to imagine a business of repeating the self-
same signifying act without end, of iterating *for ever*? Or, which will
come to the same, what would it mean to *deny* the possibility of endlessly repeating a signifying act? Is it in fact possible to coherently
imagine an activity of iterating that did not—by definition of the very
abstracted purity of the repetition that furthers it—go on forever?

Language

Observe that our contact with infinity is always and only through
writing. Certain signifiers, notations, inscriptions, and rules for operating on them—1/0 (division by zero), *lim*(1/x) or *tan*(90) (passage to
the limit), 1, 2, 3, . . . (ad infinitum), 0.3333 . . . (recurring decimals),
and so on—require or seem to require an "infinity" as their signifieds.
Evidently, one must start with mathematics' immersion in signs, nota-

tions, and symbols, its written discourse of ideograms. Without these there is (inside mathematics) no way to count and nothing to count.

How then should we regard these ideograms? The mathematical community bifurcates its discourse into a privileged formal mode and an informal one considered as supplementary and epiphenomenal. This latter mode is *unrigorous* mathematics (imprecise domains of heuristics, motivation, examples, intuition, and the like) involved in framing and scene setting and permeated by the vagueness and inexactitude of natural language. The former mode corresponds to *rigorous* mathematics (formal, exact, "proper") constructed according to officially sanctioned syntactico-grammatical rules and protocols governing acceptable and legitimate sign use.

Two crucial features of formal mathematics—call it the official *Code*—stand out. First, every text written in the Code is riddled with *imperatives*, with commands and exhortations such as "multiply items in *w*," "integrate *x*," "prove *y*," "enumerate *z*," detailing precise procedures and operations that are to be carried out. Second, the Code is completely without *indexical* expressions, those fundamental and universal elements of natural languages whereby such terms as "I," "you," "here," "this," as well as tensed verbs, tie the meaning of messages to the physical context of their utterance.

Immediately questions arise. Who are the recipients of all these imperatives? What manner of agency obeys the various injunctions to multiply, prove, consider, add, count, integrate, and so on? How is the Code's lack of indexicality related to the impersonal, transcultural nature of mathematical knowledge? Of what relevance is informal mathematical discourse—call it the supplementary Code or *meta-Code*—to formal, "proper" mathematics?

Thought Experiments

To respond to these questions one needs a conception of mathematics-as-language that reflects the dynamics of what it means to *do* mathematics—to manipulate mathematical symbols and think mathematical thoughts. To this end I have constructed a semiotic model that depicts mathematical reasoning as consisting of certain imaginings or *thought experiments* played out through written signs. These imaginings are organized in terms of three figures—semiotic agencies—that operate simultaneously at different levels of discourse.

There is the reader/writer of proper—rigorous and formally correct—mathematical texts, the mathematical *Subject* who uses the Code but has no access to any (necessarily indexical) description of itself. In this, the Subject is an agency entirely separate from the *Person*—the one who speaks informal mathematics and has full immersion in history and in the cultural subjectivity coded by the "I" of natural language that permeates the metaCode. The Subject responds only to inclusive imperatives, those of the form "let us consider, prove, define *x*," which ask that speaker and hearer set up a shared world or, a world having been set up, share an activity about it. All other imperatives such as "count the elements of *M*," "reverse all arrows in *D*," "integrate *f*," "add *p* and *q*" are exclusive: they demand that certain actions in these shared worlds be performed. Such actions are not carried out by the Subject. Instead the Subject imagines into being an idealized simulacrum of itself as its surrogate, what C. S. Peirce calls a "skeleton diagram of the self," which executes exclusive imperatives of the form "count," "integrate," and so on. I call this skeleton of the Subject the Subject's *Agent*. Basically, the Subject engages entirely with signs at the level of the Code, the Person with the signs and metasigns of the metaCode, and the Agent, an automaton without the ability to engage with any meanings, operates only with signifiers at a sub-Coded level. Operating together they enact a single thought-experimental narrative.

Such a way of looking at mathematical reasoning fits with the idea that mathematical assertions are essentially *predictions* about signs. For example, $2 + 3 = 3 + 2$ means that joining 11 to 111 will have the same outcome as joining 111 to 11; or the assertion "the square root of 2 is irrational"—that is, no ratio p/q of numbers can be squared to give 2—is the prediction that whatever integers are substituted for p and q in the expression $p^2 - 2q^2$ the result will not be 0. Thus assertions are semiotic foretellings, they predict the outcome of the Subject's future engagement with signs.

Thus, in the course of a thought experiment whose aim is to prove some assertion/prediction *X*, the Person is an observer of the Subject's imagined manipulation of the Agent and, as a result of what s/he observes, is persuaded to accept *X*. Evidently (and no differently from ordinary thought where we propel a version of ourselves around an imagined world) persuasion here will require enough likeness between Subject and Agent for the Person to believe that the Agent's imagined activities can stand proxy for what the Subject *would* encounter were the predicted situation to be experienced. We can put this another way. The thought-experimental model allows us to read mathematics as a business of making certain kinds of "rigorous fantasies" or waking dreams. The imagining Subject corresponds to the dreamer dreaming the dream, the skeleton Agent to the imago, the figure being dreamed, and the Person to the dreamer awake in the conscious subjectivity of language telling the dream. Such dreams are persuasive fictions only when (from inside language) we recognize the imago as *resembling* our imagining selves.

How close can or must such a resemblance be? The field spans an opposition. On one side, total identity: the imago *is* ourselves. But then we are acting, not dreaming about acting, performing not thought experiments but actual ones, and no contact with the problematic of imagined action occurs. On the other side, total difference: the imago can take any form and do or be anything we imagine. But then the question: why should we be persuaded its activities have relevance for us? On what *actuality* would such an arbitrarily conjured imago's thought-experimental journeys impinge?

The Ghost

In relation to the infinite, to the possibility of a never-ending counting, the resemblance between Subject and Agent becomes pivotal. And the decisive characteristic is that of physicality. Is the mathematical Agent, the imago who is to count endlessly for us, imagined to have a body—however idealized and ethereal— or is it a wholly disembodied, immaterial phantom? The answer can only be the latter: such an imago has to be something transcendental, it has to be a ghost. And it is not hard to see why. Once we grant the imago some scrap of physical being, some contact or connection, however rarefied and idealized, to the world of material process, then it will be organized under the regimes of space, time, energy, and its actions cannot be free of the effects of contingency and entropy. As a consequence, its efforts to count endlessly must fail.

And how, in any case, independently of this implication for counting, could such a no-body phantom be the source of *persuasion*? Why should an embodied mathematical Subject, whose identity and ability to interpret signs are inseparable from its physical being and contingent presence in the world, create a totally disembodied Agent as its proxy? How can the imagined doings of such a perfect, transcendental imago, possessing in Kant's phrase (for the moral absolute within us) "a life independent of animality and the whole world of sense," impinge on the Subject's inescapably corporeal presence? Of what relevance are journeys in the world of the ghost to human beings and the sensorimotor world they occupy?

The disembodied Agent then—as near to God as makes no difference—is a spirit, a ghost or angel required by classical mathematics to give meaning to "endless" counting. There is nothing in the way of argument to prevent our belief in it and, with it, the familiar embrace of the *ad infinitum* progression of whole numbers. And there is nothing to prevent us extending this to a belief in Platonic entities—in effect embedding the ghost in Plato's True Unchanging Heaven—and thence to the whole paradisal universe of infinitely many infinities that dominates twentieth-century mathematics.

Corporeality

Nor is there anything to prevent our disbelief. It is possible to reject not only Platonic orthodoxy but, more fundamentally, the very idea of disembodiment itself, to refuse altogether the imago of endlessness. In which case one imagines Agents always with suitably idealized but never absent bodies. The resulting corporealized mathematics opens out into a radically new conception of iteration, of counting, and therefore of what we might and could mean by "number." What emerges is a *non-Euclidean arithmetic* which, in its founding move, replaces the endless repetition of the orthodox picture, the iteration of the same, by an entropic diminuendo. The numbers are seen to dissipate, to fade out and become indeterminate as counting them into being is ever further prolonged—an indeterminacy that seems closer to our actual experience of iteration than any transcendentally mysterious infinitude.

Beyond such demythologizing, the employment of these numbers would have a radical impact on how we are to think mathematics' relation to the "real." Were physics, for example, to use them to measure and theorize materiality in place of the current ad infinitum numbers, the result would be a description of the physical universe strikingly different from the one currently on offer. In the absence of un-limited subdivision everything numerically measurable—space, time, mass, energy, charge, gravitation, information—would be quantized, there would be no arbitrarily-approximatable-to "points" in space or "instants" of time, all movement would be discrete and ultimately discontinuous, and so on.

But the Euclidean fantasy of pure and endless iteration and of absolute Truth has by no means been denied credence. On the contrary, as both instrument and metaphysics, it regulates contemporary technoscience and permeates our thinking about number. It is what we all wittingly or not subscribe to, what we all invoke, when we interpret the ideogram ". . ." as the injunction to go on counting without end. To recognize this, to recognize the presence of a disembodied transcendental being inside such an interpretation, is to see the fantasy *as* a fantasy. Replacing it by an insistence on the corporeality of the-one-who-counts is to start rewriting the connections between God, Number, the Body. To see mathematics, in other words, as a business with made and not found ideograms and to insist on a mathematical imago that bears some—however gossamer-like—connection to our essential contingency, empiricity, and finitude is to open up a new way of counting and a more, not less, interesting vision of "number."

. . . *1*

L
A
N
G
U
A
G
E

Philosophy is written in this grand book—the universe, which stands continually open before our gaze. But the book cannot be understood unless one first learns to comprehend the language and to read the alphabet in which it is composed. It is written in the language of mathematics.

· Galileo, *The Assayer*

However the topic is considered, the *problem of language* has never been simply one problem among others. But never as much as at present has it invaded, *as such*, the global horizon of the most diverse researches and the most heterogeneous discourses, diverse in their intention, method, and ideology.

· Jacques Derrida,
Of Grammatology

t is becoming increasingly difficult to find a region of the human sciences and humanities untouched by a certain kind of cultural/ intellectual ferment that has gathered force and energy over the past two decades. Architecture, history, anthropology, visual arts, the study of contemporary culture, literary criticism and theory, film, dance and theatre studies, musicology, philosophy: all exhibit (with varying intensity, enthusiasm, doubt, and resistance) the impact of a new movement of ideas, formulations, ways of thinking, questioning, and writing. We might call this movement the "Post-tendency" or simply the "Post" to encapsulate the way in which its adherents and proponents describe themselves as post-something-or-other (post-structuralist, post-modern, post-patriarchal, post-humanist, post-industrial, post-Enlightenment). Though of predominantly French origins, at least in its dissemination as theory rather than the socio-cultural manifestations it addresses, the movement is now widely dispersed within the Anglo-American intellectual scene. No one idea, formulation, question, cultural program, or philosophical agenda can encapsulate the shape and direction of this movement, but behind its multiform impact, operating as a key site of innovation and conflict, is the nature and business of *language*. By this I mean language in its widest possible connotation to include natural speech, writing, recording, communicating, the use of signs, codes, notations, symbolic systems, and with these all the associated issues concerning mediation, semiosis, interpretation, representation, signification, discourse, sense, and meaning.

An area of the humanities perhaps most resistant, unsympathetic to and, as yet, least affected by the destabilizing subversions of the

Post, is mainstream Anglo-American analytic philosophy. This has certain unavoidable consequences for us, since it is a discipline whose concern with and focus on logic and mathematics and the nature of scientific knowledge make it of natural relevance to the present essay. Here the refusal of the Post (essentially of post-structuralist or deconstructive theorizing) takes the form of a conflict—deep and, it would appear, irreconcilable—between an old and much-worked-through orthodox tradition regarding language and a newer philosophical outlook opposed and in many ways alien to it: a conflict between the so-called continental outlook whose modern conception of language—dominated by Nietzsche, Husserl, Heidegger, Wittgenstein, and Derrida—might be captured by the slogan "Language speaks man into the world," and the current analytic mindset—associated with Frege, Bertrand Russell, and their empiricist forebears—whose slogan might be "Man speaks language about the world."[1]

To say, then, of some human endeavor, in our case the practice of mathematics, that it is a "language" or a symbol system or a mode of discourse is to have to negotiate this conflict. Such negotiation seems inevitable if one is to place one's perception of mathematics, at this late point in the century, in relation to the intellectual current of twentieth-century Post thought. But it is clearly an engagement which brings with it certain risks and difficulties since, as will be evident from the above, the entire arena of language (and not just within philosophy) is contentious, open-ended, problematic. This entails that the description of the linguistic and signifying capacity of mathematics will have to face issues and questions unknown and foreign to its previous unlanguaged or inadequately languaged image of itself.

Leaving such problematics and negotiations aside for the moment, there is, on the face of it, little need to *insist* that mathematics is a language: who after all among those familiar with it would deny the proposition? Certainly not those users—accountants, engineers, economists, actuaries, statisticians, cliometricians, meteorologists, and the like—who have no choice but to translate in and out of mathematical expressions and terminology on their way to other interests and forms of engagement; and certainly not scientists and technologists, the

professional practitioners of the physical and life sciences and technologies who have always appealed to and depended on mathematics as
an essential linguistico-cognitive resource. Is not the contemporary
technoscientific picture of physical reality literally unthinkable outside
the apparatus of mathematical notations and terms used to articulate it?
For Galileo, reading these terms was not so much articulating the language of science as deciphering the writing of God; and if in the four
centuries since his proclamation science has progressively sought by a
series of disclaimers to distance itself from the more blatantly intrusive
aspects of Galileo's theism, it has left his understanding of mathematics
as the language and alphabet of the physical universe very much in
place, adhered to with remarkably little questioning by innumerable
scientists from the seventeenth century to the present day.[2]

Yet, despite the many and much-repeated recognitions of the linguistic nature of mathematics, there has been little sustained attempt
to develop the philosophical and conceptual consequences of saying
what it means for mathematics to be a language or be practiced as
a mode of discourse. A notable and important exception would be
Wittgenstein's extended remarks on the foundations of mathematics
with their intent to characterize the doing of mathematics (at least in
its elementary computational aspects) as a motley of certain kinds
of language games. But, though sharp, interesting, and unfailingly
provocative, Wittgenstein's fragmentary and idiosyncratically unsys-
tematic dicta do not address the *theoretical* question of mathematical
language and discourse in any direct or even indirect way. Nor do
they indicate how one might do so. Notwithstanding this, however,
much of what follows here is sympathetic to his overall philosophical
enterprise, notably the radical nominalism running through his critique of Platonistic metaphysics which sees any mathematical "object"
as an effect of the notation system that supposedly describes it, and
(less immediately) the sorts of worries about our understanding of
number and following the rule of counting that he so persistently
puzzled over.

Galileo's amalgamation of language, writing, books, and alphabetic letters raises certain crucial issues: what do languages—divine,

natural, or man-made—have to do with alphabets? Is mathematics a "language" or a form of writing—is there indeed a difference? And related to these: is there a distinction to be drawn between "depth" and "surface," between, for example, a project for a "grammar" of science and the apparently less objective, more impressionistic study suggested by "language" of science? And why *of science*? Why not the grammar, the language, the code, the symbolic system, or whatever, of mathematics itself—cut free from all questions of its instrumentality? We shall return to these questions later.

For the moment the original puzzle remains: why has the widespread recognition of mathematics-as-language not given rise to any theorization of the nature of this language or of the status of mathematics as a discursive practice? Any answer must start from the fact that for mathematicians and many others the predominant conception of mathematical language—the one that enters into these recognitions—comes built into and inseparable from a certain account of mathematical "reality" they take it to relate to. The account, called variously "realism" "metaphysical realism," or Platonism, goes briefly as follows.[3] One starts from the assumption of a specifiable ontology— that the world consists of a definite totality of discourse-independent objects and properties. In relation to these objects one assumes a bivalent logic of excluded middle—each object either has or has not any property meaningfully applicable to it. One then operates with a truth-based epistemology—the world is described by language through the principle of correspondence, in which properties correspond to collections of objects, and statements are either true or false, depending on whether the states of affairs they correspond to are "really" the case—that is, whether these states either do or don't obtain in the world.

In mathematics this Platonistic doctrine, though in some version or other as old as its name suggests, was given its modern formulation at the end of the nineteenth century: directly by Frege in his attack on any psychologistic reading of mathematics which would include the mathematical "subject" and in his reduction of mathematical entities to logical objects and concepts situated outside of human activity;

and indirectly through Dedekind's and Cantor's introduction of infinite sets as well-defined mathematical objects in terms of which number and real magnitudes were to be unambiguously characterized. Despite numerous arguments, critiques, ridiculings, and attempted refutations throughout this century, metaphysical realism still stands, subscribed to and defended by many philosophers, and held tenaciously by the majority of mathematicians for whom it is the prevailing orthodoxy. Our purpose here is not to add to these philosophical critiques, but only to identify Platonism as an obstacle to any adequate characterization of mathematics as a language. For this exercise the key element of the doctrine lies in its founding ontological assumption of a *discourse-independent* world, a world of mathematical objects—numbers, points, lines, sets, functions, morphisms, spaces, and the like—that are held to exist prior to and independent of any talk, description, or discussion of them. According to this assumption, mathematical languages refer to and make sense of these objects and generate assertions about their properties and interconnections, but in no sense can any aspect of mathematical discourse impinge on their constitution, origin, nature, properties, or existential status. For contemporary apologists of mathematical Platonism (especially, but by no means exclusively, mathematicians) the belief in objects "out there"—uncorrupted by the vagaries and uncertainty of history, culture, human choice, and the assorted subjectivities that permeate discourse—is crucial and nonnegotiable, any concession appearing as a recapitulation to some form of psychologism or relativism.[4]

What is at stake here in these and numerous similar refusals to compromise on, or justify, or even attempt to explicate the "out-thereness" of mathematical objects is no less than the principle of absolute Truth—truth transcendentally removed from any possibility of human agency, truth perfect, truth outside time, outside space, outside the material world of matter, energy, and entropy, truth somehow able to be discovered or apprehended but never invented, discernible but in no way decidable by human intellect. And with this truth as total adequation, perfect item-by-item equivalence between human assertions and affairs in some supposed prior world, put into place via a principle of correspondence that is as circularly absolute as that which it makes correspond, comes, as both support and consequence, the classical framework of knowledge. Thus, within this horizon, there is first ontology and being, the inventory of the objects that are or must be in this already-given world; then reference, pointing, and naming whereby language, in an activity external to and after-the-facts of this world, picks out these preexistent objects; from reference comes sense, the description and meaning of the properties and states of affairs enjoyed by these objects; finally epistemology, the examination of the means of knowing, believing, validating which among the assertions generated by language about these objects is a justifiably "true" description of the states of affairs they take part in.

The order of events here—being, referring, meaning, knowing—is a crucial element in the way this framework of what constitutes "knowledge" works to bolster the metaphysics of Platonism. Thus, to suggest, granting the terms themselves but not Frege's account of them, that sense might precede or even be detachable from reference, that what a language is "about" might not be separable from the discursive employment of that very language, is to suggest the possibility of usurping the priority of "things" to names, of inverting the order of being and knowing by raising the specter of language affecting, modifying, and indeed *creating* what it talks about. And to do this is to start pulling apart the classical notion of representation according to which language represents when it re-presents; representing by presenting again as secondary what was prior, originally given as present, determined, and existent outside and before language. Much, then, is at stake and much at issue—the dissolution of the sense/reference opposition, the abolition of truth as adequation, the denial of the anteriority of "things" to names—in any breaking free from the rigid separation and subordination of a posterior language to some prior reality "out there" insisted upon by any kind of metaphysical realism. In the light of this sort of collusion between the present-day, post-Fregean epistemological framework and the founding beliefs of mathematical Platonism, it is no surprise, then, that the dominant mode of discussing mathematics, certainly within analytic philosophy, is not as a language or as a discursive practice but as a species of *knowledge.*[5]

In short, the understanding of mathematics-as-language from within Platonism—either embraced nakedly by mathematicians as the refusal to give up on the idea of timeless mathematical truths or fully dressed with any number of philosophical refinements, layerings, and caveats as a commitment to this same truth in the guise of suitably founded knowledge—is to see language as a powerfully useful but inert medium; a pure vehicle for re-presenting a prior world whose objects owe nothing to the nature, use, or presence of the medium and whose function is to do no more than descriptively reproduce, with neutral and transparent efficiency, the true features possessed by these objects. But a conception of mathematics as a species of knowledge and of language as the transparent medium for conveying and transmitting it has become, whatever insights and energy it might once have provided in the Cartesian tradition, explanatorily inert and passive. And it is precisely because nothing new issues from it—about mathematical practice or the constitution of its objects—that the conception is so easy to own and assent to as an unproblematic and obvious truism: hence the many recognitions, readily given and in good faith, that mathematics is a "language"; and hence also the fact that such acknowledgments don't—in employing this understanding of language—amount to very much. In most cases they are, indeed, little more than a recognition of the extended symbolicity of the discipline, the evident capacity of mathematics to introduce symbols and ciphers and engage with hieroglyphic-like inscriptions and notations via arrays of formal rules.

Evidently, the means for articulating what is to be understood by mathematics-as-language will not come from within the self-confirming circle of a transparent medium conveying "knowledge" about timeless objects that Platonism moves in. And since no sort of philosophical critique of Platonism's claims seems capable of making an impact—too many have already been given and ignored (at least by the majority of mathematicians and their influencees)—only some form of direct action, it would seem, could break out of the circle:

some way of *doing* mathematics that repudiated, as an integral part of its practice, at least one of the interlocking assumptions that make up the apparatus of metaphysical realism.

Early in this century precisely this happened through Jan Brouwer's formulation and prosecution of the intuitionist program, with its outright rejection of classical logic and its insistence that mathematics was an activity of making mental constructions. Why, Brouwer asked, should we accept the principle of bivalence, the classical law of excluded middle that demands of any mathematical assertion, *P*, that either *P* holds or not-*P* holds? If mathematical existence claims have to be backed by constructions rather than the logic of contradiction, and if proofs are the means of specifying how these constructions are or could be carried out, then it might and does happen that for many cases we are not (and may never be) in a position to specify a construction for *P* or specify one for not-*P*, that is, a construction which would yield an absurdity from the supposition of *P*. Brouwer's "constructions in the mind" put onto the mathematical scene an alternative to the conception of mathematical assertions as descriptions, and proofs as their validations, of prior existence. Since then other forms of constructivism have been developed with the result that there is now a significant minority of mathematicians who, despite disagreements among themselves, dissent—via their practice—from the manner of doing and interpreting mathematics countenanced by Platonist orthodoxy.[6] For our purposes one can ask therefore whether such dissent—by removing the obstacle put in place by Platonism—forms the appropriate sort of platform from which to mount an articulation of mathematics-as-language.

The answer, as we shall see, is (a much qualified) yes: the linguistic or semiotic model of mathematics to be presented later in this essay rests fundamentally on construing mathematics as a species of mental activity or construction, specifically as a certain kind of thought experiment. In this sense the account of mathematics offered here is undeniably a constructivist one (though a great deal of explanation of the meaning of "construction" needs to be registered before such an affirmation can be accepted; and more importantly, in relation to the present essay's interest in the status of infinity, an analysis of the way the existing forms of constructivism theorize—or, rather, fail to theorize—the integers, needs to be worked through). For the moment let us observe that in the case of intuitionism, certainly as presented by Brouwer, its capacity to be compatible with (let alone engender) a linguistic view of mathematics is entirely problematic. This is because for Brouwer mental constructions were private activities carried out in our inner intuition of time: mathematics is shared or common to us all, not through discourse, not through any circuit of communication or exchange via the public or intersubjective agency of signs, but because each of us, each and every individual mind, operates within the same Kantian categories. Thus, what I intuit so must you, regardless of how you or I might express, describe, symbolize, or articulate such intuitings. "Mathematics," Brouwer infamously insisted, "is a languageless activity." The *practice*, then, of constructivism, though responsible for effecting a radical break with the metaphysical axiomatics of Platonistic truth and existence, does not in itself provide a persuasive basis for setting these axiomatics aside.

Nor, more fundamentally, is it self-critical enough, as we shall see, to allow one to discern behind Platonistic metaphysics a less obvious and altogether more subtle metaphysical principle—the transcendentally constructed ghost of this essay's title—at work inside our picture of the integers. Nor, nearer to the present point, does the practice of mathematics by constructivists, unaccompanied as it is by any general account of what goes on when mathematical signs are manipulated, engage with any questions concerning the linguistic apparatus and discursive moves that facilitate their own activities. Indeed, as it stands, the constructivist replacement of classical epistemology's items of "knowledge" by constructions, and of bivalent "truth" by an alternative logic of justification, forms only a partial renunciation of the framework within which this epistemology operates. It leaves the question of the *being*, or rather the becoming, of basic mathematical objects such as numbers and points virtually untouched.

But, as will be evident by now, outside a metaphysically presented realism, the ontological question of being and existence cannot be separated from the question of language. The two come directly together in the problem of "reference": once one can no longer rely, as Platonism so magically does, on a domain of objects in place before the advent of language, waiting to be named and pointed to by linguistic items independent of them, then what, if anything, is one to mean by a referent? Such an external pairing of posterior names and objects priorly innocent of them becomes unsustainable once the possibility arises that the objects supposedly pointed to by language are themselves shaped, facilitated, evoked, created by the same discursive means used to "name" them. There is, then, a sense—a very definite, explicable, and important sense to which we shall return—in which mathematical language and discourse deal in, are oriented toward, and are "about" mathematicians' own inscriptional activities; so that, if one insists on using the term, mathematics might be said to "refer" (like music) to nothing other than itself.

This is not to say however that such an endomorphic characterization of mathematics' relation to its object exhausts its character or status as a language. Mathematics is neither a self-contained linguistic formalism nor an abstract game played entirely within the orbit of its own self-referring rules, conventions, and symbolic protocols. On the contrary, mathematical discourse would rapidly become unintelligible as a signifying practice and would never be possible as a form of instrumental reason were it not for its relation to some—any—"natural," nonmathematical host language (English, Urdu, Latin, Chinese) through, against, and in proximity to which it is certainly de facto and, as I shall show, de jure practiced and interpreted as well as used and applied. Observe that mathematics, being (as a result of a long, consciously pursued program of rigor and objectivity) trans- or pancultural, can bear the imprint of no particular language either in its content or its procedures. This means that what is being implied here is that the signifying capacity—meaning, persuasiveness, and intelligibility—of mathematical signs requires certain minimum linguistic means, certain properties of natural languages, to be present (though not acknowledged or even perceived as such) in every act of mathematical thought. Of particular interest will be that apparatus of signs which allows an embodied subject—the corporeal, situated speaker of natural language—to register a presence in and connection to the world of real time, space, and physical process.[7]

Let us return to the question of "language," of how we are to think mathematics-as-language. Despite any claims for the essential intimacy and interdependence between mathematical practice and some host language, there is no danger of confusing the two, and certainly no warrant for trying to conceptualize mathematical language as if it were a natural language. Unlike the latter, mathematics is before all else self-consciously produced; and it is so according to an agenda formed out of its historically conditioned role: as instrument in relation to the needs of both commerce and technoscience and, with greater autonomy, out of the image of itself as the exercise and play of pure, abstract reason engaged in the production of indubitable truths.

But a more immediate separation of the mathematical from the natural lies in the all-important fact that mathematics is a particular kind of *written* discourse; a business of making and remaking permanent inscriptions—symbols, figures, notations, graphs, marks, diagrams, equations—written down on paper, blackboards, and screens and manipulated—that is, operated upon, transformed, indexed, amalgamated, arrayed, rearranged, juxtaposed, sequenced, and ordered—according to a vast, highly developed and complex body of rules and procedures. The sense of writing here should be contrasted at the outset with the conventional and standard conception of "writing" in Western culture that derives from the use of the alphabet. According to this, writing is essentially a secondary phenomenon, a recording or transcribing of the primary business of speech: what is written allows the reproduction of the sounds of spoken words, and interpreting writing is no more than a transposed version of hearing the spoken code that precedes it. Such is not the case with mathematics. In contrast to the secondarity ruling alphabetic writing, mathematical signs do not code, record, or transcribe anything extramathematical: mathematical items evoke and mean what they mean, what they are to signify, directly and not as intermediates for something else.[8]

In short, mathematical signs are to be understood as *ideograms* in the usual sense of written characters conveying, invoking, or denoting conceptual content—signifying—through their graphic identity, as visually presented marks. Of course, an ideogram—"1," say—can be spoken and named in speech—"one," "un," "eins" . . . —but its manner of functioning in mathematical discourse, the meanings it creates, the forms of significant mathematical content it facilitates and gives rise to, is configured internally. It operates not in terms of all these different spoken names but through its conformity to the rules, arithmetical protocols and narrative presuppositions governing the signs for "number," that is, governing "1" and all the other written number characters "2," "3," . . . that come after it. Of course, there has been over their entire history a dialectical movement between the arithmetical signs "1," "2," "3," . . . and the natural language words such as "one," "two," "three," etc., resulting in a motivation and concretization of the arithmetical signs from one side and from the other in the colonization by mathematics of natural language's capacity to enumerate. But this mutual entanglement of "1" and "one," the basis for any account of the motive and utility of arithmetic, in no way detracts from the ideogrammatic status of "1" as a self-standing arithmetical character. Similar remarks apply to many other characters marking historically rich sites of interpretation and networks of meanings which circulate throughout mathematics. The ideogram + of "addition" (of numbers, operations, functions, spaces, processes), for example, becomes the ideogram $\sum$ of finite summation and then by way of a certain principled and conditioned repetition the central ideogram $\int$ of the integral calculus. On a different route + gives rise to the ideogram δx meaning a small addition to an independent variable, which in turn leads to the founding ideograms dx and ∂x of the differential calculus.

But identifying the forms of an ideogram and relating ideograms either through their historical development or synchronically, structuralist fashion, to each other, though interesting and necessary for an understanding of their interconnectivity and scope, only goes so far. It does not impinge on the more primitive and originary signifying level

where the ideograms = and "0" and ". . ." which underpin the arithmetical progression 0, 1, 2, 3, . . . in relation to which all varieties of
+, for example, have their primary field of operation. And in the case
of these founding ideograms, substantive questions remain outside
the scope of such relationships. Thus, the meaning, purport, sense,
significance, interpretation, and application of the progression 0, 1,
2, . . . of integers and the arithmetical apparatus they give rise to rides
on the intended signification of the ideogram ". . .". And it is the
whole purpose of the present essay to interrogate the status of this
ideogram by asking what assumptions are in play and what manner of
sense and purpose there is in interpreting it as the injunction "go on
forever" or, if there is any difference, "go on without end."

If mathematics is a mode of writing not to be confused with the
transcriptional, alphabetic writing of speech and an artificially created
language manifestly different from natural language, then we are presented with an obvious and immediate question of *how* "unnatural":
in what ways do its texts produce meanings that differ fundamentally,
in kind and scope, from those of speech and standardly conceived
writing? How, for example, are its reasonings and proofs related to
and divergent from written narratives and the arguments and dia-
logues of ordinary discourse? This in turn leads to the question of
method: what vocabulary of description and explication are we to use
in order to articulate the form and manner of mathematical activity?
The latter question is, of course, methodologically primary; it asks
how and where we are to begin. Now, whatever else might be con-
tended, questioned, made problematic, or held in abeyance about the
doing of mathematics, its practice as the putting into play of ideograms, as the manipulation of written *signs*, can hardly be denied.
And sign systems—any kind of signs, whether dance notations, words,
medical symptoms, hand gestures, symbols, . . . —are precisely what
semiotics, professing itself to be the general study of signification, is in
the business of investigating.

Put forward as a proposal for a new science, a program for studying "what constitutes signs, what laws govern them" by Ferdinand
de Saussure, a Swiss structural linguist, but never in fact consummated

by him, and as part of a much larger and differently motivated taxonomy and science of signs worked out in great detail by C. S. Peirce, an American pragmatist/idealist philosopher and mathematician, semiotics is a discipline with a double lineage that offers different definitions of "sign" embedded in divergent and what would seem to be irreconcilable modes of approach.

For Saussure a sign is a coupling of a materially based *signifier* (word sound, flag, gesture, and so on) and a nonmaterial, mentally based *signified* (word meaning, thought, idea). The coupling is supposed arbitrary and unmotivated: so that, in the case of words, the signified "doghood" is coupled with the signifiers "dog" or "chien" or "Hund," not through any natural or motivated connection between idea and sound (the very multiplicity of signifiers indicates this), but through the contingencies of social convention. Not only are signs internally unmotivated in this sense, but they are also, for Saussure, what we might call externally unmotivated: signs get their sense, Saussure insisted, not through any "positivity," that is, any connection via reference, ostension, or pointing to the world, but through their relationships of difference—exclusion, opposition, contiguity—to other signs within the same linguistic or semiotic system. The insistence that language is constituted out of pure differences and characterized in terms of abstract, internal structure is codified in Saussure's celebrated structuralist proclamation that "in language there are only differences *without positive terms*. Whether we take the signified or the signifier, language has neither ideas nor sounds that existed before the linguistic system, but only conceptual and phonic differences that have issued from the system" (1966: 118–19). Of course, as Saussure had no wish to deny, signs can be and are hooked up to the world of things and events—"doghood" to the familiar four-footed particulars—but this linkage, even if one considers it to be their reason for being, is for Saussure a matter of the *use* and discursive employment of signs, not of their underlying, constitutive character as elements of a system. Saussure maintained this distinction through his strict separation between the underlying abstract system *langue* (language, code) and the actual employment *parole* (speech, discourse) of this system.

Peirce's definition is richer and more ambiguous: "A sign . . . is something which stands to somebody for something in some respect or capacity. It addresses somebody, that is, creates in the mind of that person an equivalent . . . or perhaps more developed sign. The sign which it creates I call the *interpretant* of the first sign. The sign stands for something, its *object* . . . not in all respects, but in reference to a sort of idea, which I have sometimes called the *ground*" (1931–58, 1: 135). According to Peirce three types of sign use or significative function must be distinguished, corresponding to three radically different kinds of coupling between sign and object. These are the symbolic (conventional coupling, the result of a "law"—the condition of most words and essentially of all signs in Saussure's scheme), the iconic (coupling motivated by a homology or likeness—diagrams, onomatopoeia), and the indexical (signified depends on the physical context of sign use or origination—photographs, the word "here"). Observe that nothing in Peirce's formulation prevents the interpretant of one sign being itself a sign, which in turn will have an interpretant, thus allowing the possibility (made much of by Peirce and, in their own ways, by others such as Roland Barthes, Umberto Eco, and Jacques Derrida) of a so-called infinite semiosis, an unlimited spread of signification, from a single sign.

Saussure's signifier is not to be confused with its material substrate, its physical presentation as "sign-vehicle." Thus, though Saussure slips between talking of signifiers as being "sounds," "sound-images," "sound-forms," or "phonic substance," his usage, particularly when he considers the phonetic differences language countenances, seems to avoid the identification of a signifier with its concrete manifestation. Again, while Peirce's formulation refers to a sign as "some*thing* which stands . . ." he is in no doubt that that "thing" is not to be taken as the sensible particular which is experienced. Indeed, for Peirce (who coined the terms) a sign is not a *token* (concrete, spatio-temporal occurrence—a particular "the" imprinted on this page, for example) but a *type* (law or concept determining whether a physical occurrence counts as an instantiating token—the form of the definite article in English).

A parallel and more subtle version of this distinction occurs in relation to the "subject"—the utterer, addressee, sign-user—the "somebody" for whom or to whom, according to Peirce's formulation, the sign stands. Among the sign uses Peirce identifies as indexical one would include the terms "I," "you," "here," "now," "this," as well as other demonstratives and tensed verbs, all of which tie the utterance or message to the physical circumstances—the space, time, place—of its utterance. Such terms, named variously by writers after Peirce as "shifters," "deictic elements," and "egocentric particulars," have been recognized as fundamental constituents of any natural language and by some, notably the linguist Emile Benveniste (and in a psychoanalytic direction by Jacques Lacan), as an essential part of the means by which human subjectivity itself gets constituted. Thus, according to Benveniste:

> Language is . . . the possibility of subjectivity because it always contains the linguistic forms appropriate to the expression of subjectivity, and discourse provokes the emergence of subjectivity because it consists of discrete instances. In some way language puts forth "empty" forms which each speaker, in the exercise of discourse, appropriates to himself and which he relates to his "person," at the same time defining himself as *I* and a partner as *you*. (1971: 227)

Under this reading of the first-person-singular pronoun, the individual speaker who says "I" is an instantiating token of the type—the empty form "I"—made available by language. Engaging this type, employing it in discourse, creates the—abstract, semiotic—*subject*, an entity distinct from the user who tokens and materializes it, since it is the type which activates and confers upon the user a subjectivity—constituted in relation to an "other," a "you"—perceived as an autonomously ongoing "self" but at the same time bound by the structure of language and the protocols of the discourse in which it operates and declares itself. Though Peirce never discussed indexicals in this way and did not attempt any corresponding theorization of the "subject," he was much alive to the materializing proclivity—the inseparability, that is, of consciousness from sign, of subjectivity from signifier—motivating it. "The word or sign which man uses," he wrote, "*is* the man himself"—or herself, of course. And again, prefiguring Wittgenstein's declaration in his *Tractatus* that "the limits of my language mean the limits of my world" (1961, §5.6), as well as Heidegger's understanding that language speaks Man, Peirce declared "Man is a sign. . . . Thus my language is the sum total of myself" (1931–58, 5: 189).

Not surprisingly, given the degree of abstraction and scope of semiotics' ambition to be a science of any kind of signification, including natural language, the characterizations of "sign" and associated conceptions of "language" presented here have been the subject of intensive questioning. This has ranged from attempts by semioticians to unify the manifestly different definitions of "sign" derived from Saussure and Peirce—or to declare them irreconcilable—through criticisms of the misplaced positivity inherent in any goal to produce a "science" of signs, to the more radical intent of exposing and deconstructing the metaphysical assumptions that thread through Saussure's and Peirce's writings, particularly the coherence of the notions of "structure," "interpretation," "object," "communication." These critical moves cannot be elaborated without going well outside the remit of the present essay: my interest in semiotics here is not, after all, its methodological foundations, its aspirations to scientific status, or the supposed metaphysics of its formulations, but—and the reduction is undeniably simplifying and opportunistic—almost exclusively its use as a possible scaffold, a working formalism ultimately to be put aside, with respect to which one can at least initiate the project of articulating the nature of mathematics-as-language. And for such a purpose, all that is required is a certain vigilance, an awareness of the difficulties that lie behind the basic semiotic vocabulary—which is principally Saussure's, but was opened up by Peirce's richer understanding of signifying[9]—of *sign*, *signifier*, *signified*, *language* or *code*, *discourse*, *metasign*, and *subject* that we shall employ. With this in mind we can make the following observations concerning some of the sites where this vigilance might be necessary.

There is the whole question of the supposed referent of a sign—

the issue, that is, of the assumed "aboutness" of signification. From Saussure's structuralist perspective the question does not arise in any interesting way, since the relation between signs and some putative reality is for him a function not of language but of speech and its uses, it belongs to the domain of discourse—secondary and subordinate to the systematic features of the code—and is accordingly not of primary concern. The issue, however, can hardly be left there, particularly if one disputes, as we shall later, the relation between code and discourse that Saussure operates with. Quite the opposite attitude to a sign's referring is the case with Peirce: the aboutness of any sign is part of its definition as a thing "which stands to somebody for *something*." Peirce calls this something the signifier's "object." Unfortunately, despite his many attempts to explicate the notion, what he means by it is obscure and unresolved, and the obscurity seems to go deep within his thought. The problem is Peirce's conception of "reality"—within which "objects" are located—whose philosophical apparatus insisted on separating all phenomena and modes of being into categories: "firstness" (possibility, experience "in itself"), "secondness" (brute actuality), "thirdness" (habit, prediction). Within this strange and unstable amalgam of Kantian idealism and hard-nosed pragmatic realism, Peirce's idea of an "object" seems to oscillate between mutually contradictory formulations.[10]

The issue of referentiality cannot be left there, particularly in relation to the present project. For it is precisely the question of the status of mathematical "objects"—the "aboutness" of mathematical signs, the nature of mathematical "reality"—that one wants semiotics to illuminate. And Kantianism, however mediated, seems able to convey nothing new on this score, nothing that is not already part of the picture of mathematical "objects" as mental constructions offered by the intuitionists and their successors. This is not to say, however, that Peirce's description of how we think these "objects" (his pragmatic understanding of them from a mathematician's viewpoint as opposed to their derivation within his system of metaphysics) is without value. On the contrary: we shall later rewrite his analogy between mathematical reasoning and the structure of everyday thought experiments as the founding abstraction of our semiotic model.

In fact, what is distinctive about mathematical constructions, the feature which takes them outside the orbit of any idealized interiority, whether Kantian transcendentalism or any other purely mentalistic discussion, is their relation to *writing*: to do mathematics is to be involved in the corporeal practice of making physical inscriptions, in which signifiers given meaning, point, application, and motive in relation to certain objects such as numbers, lines, or morphisms being signified are read and scribbled. The nature of the mathematical sign, the coupling of signifier and signified, *is* the relationship between all this scribbling and thinking, between what the body writes and what the mind "constructs" and imagines. Now because writing is assumed—on the alphabetic, transcriptional model—to be an activity secondary to speech, and because in the history of Western culture the mind has been valorized above the activities of the body, there is a natural historical pressure operating to impose this double hierarchy here. The effect of this is to rank mathematical thinking as more important than, primary and superior to, mathematical writing, to hold that mathematics is signified-driven: first meanings then notations, objects then names, ideas then expressions, numbers then numerals.

But such, as we shall demonstrate, is not the case: on the contrary, no account of mathematical practice that ignores the signifier-driven aspects of that activity can be acceptable. It is simply not plausible—either historically or conceptually—to ignore the way notational systems, structures and assignments of names, syntactical rules, diagrams, and modes of representation are constitutive of the very "prior" signifieds they are supposedly describing. The upshot will be that symbol and idea, writing and thinking, signifier and signified have to be understood as co-creative and mutually originative. Certainly, mathematicians write down what they take to be prior meanings, but they can only realize and construct such intersubjective signifieds in relation to written signifiers already sanctioned by mathematical discourse, which in turn can only be thought in relation to prior signifieds, . . .

We began by asking how we are to think mathematics-as-language and have ended by arguing that the only place to start is with the undeniable fact of mathematics being a manipulation of written symbols. The path we have followed has gone from an outright denial of Platonism's capacity to provide any kind of linguistic understanding (language as transparent medium; sign under the regime of timeless transcendental "truth") through a dissatisfaction with Kantian constructivism (language as epiphenomenal; sign as mere secondary effect of mental activity) to a yet-to-be-explicated version of what one might call semioticized constructivism (language as constitutive activity; sign as a species of thought-writing). Now semiotics, as I have indicated, rests on a series of highly questionable binary distinctions in the form of oppositions, principally signifier/signified, discourse/language, object/interpretant, but also structure/genesis and arbitrary/motivated, in which one of the pair is given primacy over the other. In the case of mathematical signs I have claimed, and have yet to justify, that each of the principal hierarchicalized oppositions—signified over signifier, language over discourse, referent over sense—must be first overturned, then reinterpreted so as to abolish any sense of primacy as well as any sense of absolute separation between their terms.

Just such an overturning of these and other cognate oppositions, within the human sciences and not in relation to the kind of project for mathematics envisioned here, has already been announced and worked through during the last two decades as part of a large-scale reaction against structuralist thought. The main thrust of this has been against structuralism's propensity to freeze what it studies into over-schematized, motionless diagrams. This reaction has been in the form of an insistence that, far from being theoretically capturable or thinkable in terms of static structures, what is significant and vital about all human activity is its local and open-ended temporality, the way it is embedded in and inseparable from *process*. This means that only by understanding structure in terms of its genesis and theorizing genesis as the movement of structure can a disabling reduction of human complexity be avoided.[11] One outcome of this insistence on the mutual entanglement of structure/genesis or state/process in relation to human signification is to reinstitute the importance of argument and rhetoric—the *process* and dynamics of reasoning—at the expense of logic—the *structure* and laws of reasoning—which, at least since the Renaissance, has been given dominance.

For mathematics the consequences of overturning this opposition would be momentous. To argue, as indeed we shall, that mathematical reasoning—purest and most distilled application of the timeless structures of logic—is to be seen, not as a form of inference engaged in preserving eternal truths, but as an historically conditioned species of rhetoric, as an activity whose business is always to achieve persuasion, is to push a post-structuralist understanding of signs into territory institutionally and conceptually resistant to it. In particular, such a program would entail a rewriting of the history of mathematics. Thus, instead of a history admitted to be contingent and open-ended on the level of human institutions but denied these features at the level of its putatively timeless content, one would have to insert history, and hence contingency, into the concepts of mathematics itself. This means insisting on mathematics as a wholly historical product, according to which every single aspect and item of it—*including the whole numbers*—would have to be created. In the light of this the numbers, far from being timeless, "natural," before us, transcendental, and (God-)given, would appear as time-bound human constructions which, more interestingly and more precariously than in their simple theistically guaranteed certitude, exist entirely as objects of discourse: created, controlled, manipulated, maintained, persisting and transmitted through time by that vast and dynamic apparatus of signs that we call mathematics. But if numbers are created and written in history, they can be uncreated, rewritten, deconstructed, altered by the same processes of writing and imagining that engendered them.

In that case the status of their supposed "endlessness" changes from that of an attribute of something prior to them, a "fact" describing some atemporal reality, to that of an interpretive practice, a way of writing, using, reading signs. Infinity, in other words, becomes inseparable from certain effects of the signifier, a phenomenon of mathematical texts, grammar, syntax, notations, and discourse. And as always with any interpretation-based activity, with any way of reading texts, however sanctioned and time-honored, there is the

question: why not another way? Of course, mathematics, more perhaps than any human discourse, tries to foreclose on alternative,
unintended interpretations of its fundamental signifieds, tries to ren-
der them, through the very forms of its argumentation, unthinkable,
unimaginable, and inconceivable. In this it cannot succeed . . . but
this is to leap way ahead of ourselves.

. . . 2

MATHEMATICAL INFINITY / FINITY

Here Sarsi . . . does his best to show me a very poor logician for my having called a certain enlargement "infinite.". . . Sarsi has indeed a large field here for showing himself a better logician than all the authors in the world, among whom I assure him that he will find the word "infinite" chosen nine times out of ten in preference to "extremely large."

· Galileo, *The Assayer*

The finite disappears before infinity and becomes zero (un pur néant).

· Pascal, *Pensées*, para. 223

God made the integers, all the rest is the work of Man.

· Leopold Kronecker

"Ought the word *infinite* to be avoided in mathematics?" Yes; where it appears to confer a meaning upon the calculus; instead of getting one from it.

· Ludwig Wittgenstein, *Remarks on the Foundations of Mathematics* t is surely impossible to think of the infinite and the finite independently of each other. Infinity is that which cannot be traversed; it indicates the passage to the limit, the movement of transcending, of going beyond, of overcoming and nullifying the here-and-now of the finite. And the finite? What else "is" it, how else name and conceive it, except as that which is given an identity by being refused, negated by, set against the non- or un- or in-finite?[12]

Evidently, the opposition finite/infinite has impinged on mathematics since its origination as the abstract study of quantity and length. Once numbers and lines are conceptualized as idealized entities, the question of limits—"how many numbers are or can there be" and "how long is or can a line be"—will be both immediate and unavoidable, and responses to the question become part of the founding intentions of arithmetic and geometry, making the history of mathematics and the history of infinity inseparable. The dynamics of this relationship between mathematical practice and the conceptual or ontological limits of its objects has been neither even nor continuous. There have been three particular moments in the history of Western mathematics when the question of infinity—what it means and how we are to think it—has intrusively pushed itself to the forefront of mathematical consciousness and, by being the source of inconsistencies, contradictions, paradoxes, antinomies, and other productions of discourse intolerable to mathematical reasoning, has demanded, if not a "solution," then at the least a recognition and questioning of its presence and status.

The first, virtually at the origin of mathematics and Western deductive thought, was the crisis brought into existence by the infamous paradoxes of Zeno. A disciple of Parmenides, Zeno sought to defend his master's vision of a static reality, an unchanging, eternal, unbroken One behind the world of mobile, plural, and ever-changing appearances. Thus if, contrary to Parmenides' denials, either space or time were really pluralities, then, so Zeno argued, they must be made up of atomic units; this means that either they are divisible without limit or there exist limits of subdivision. Depending on which hypothetical alternative one chose, Zeno responded with an argument that reduced one's choice to absurdity. For example, if one assumed the first, then the paradox of never reaching a destination ensued:

> If a man is to walk a distance of one mile, he must first walk half
> . the distance or one-half a mile, then he must walk half of what
> remains or one-fourth of a mile, then again half of what remains,
> ad infinitum. An infinite series of finite distances must be succes-
> sively traversed if the man is to reach one mile. But an infinite
> series is by definition a series that cannot be exhausted, for it
> never comes to an end. Hence the man can never reach the end
> of the mile, and seeing that the same argument can be applied
> mutatis mutandis to any finite distance whatever, it is clearly
> impossible for motion to occur.[13]

Zeno gave similar paradoxes, such as that of the stadium, the arrow in flight, and so on, to cover the other possible responses open to his interlocutor.

In relation to the present account Zeno's reductions of plurality and motion to paradoxical absurdity had two major and long-lasting consequences. First, there was the effect on Greek mathematical thought itself. Here, because of the dominance of the spatio-visual over the computational and symbolic that was such a feature of that mathematics, it was registered in geometry as a distrust of motion, which disallowed proofs that made reference to a moving point (an avoidance echoed within the entirely de-temporalized and static pre-sent-day conception of mathematical objects), and as a distrust of the idea of an infinitely extended straight line, which prompted Euclid's unease over the status of his axiom about parallel lines—a concern enlarged and, as it were, finally vindicated by the formulation of non-

Euclidean geometries in the nineteenth century. And then, more explicitly and over a wider philosophical terrain, there was the legacy of Aristotle's treatment of the paradoxes (long after their living impact). For Aristotle the way out of Zeno's absurdities was to accept the infinite subdivision of space and time, provided that one interpreted the notion of "infinity" used in such subdivisions as the *potential* infinity, the infinity of an endless coming-into-being, as against any conception of an *actual* or completed infinity that could be grasped in its entirety as a single object. And though, as is well known, the Aristotelian treatment of motion (particularly as this occurs in the arrow paradox) had to be overcome before Galileo could formulate the notion of instantaneous velocity so crucial to his (and subsequent) physics, the tension between a potential infinity-as-process and an actual infinity-as-object has never ceased to operate within mathematics. And, indeed, Gauss famously repudiated the latter, claiming that only the infinite as a never-ending process had any place within a properly regulated mathematical science—a repudiation and claim repeated, as we shall see, with little change by constructivists from Kronecker to his present-day heirs.

The second intrusion of infinity into mathematical consciousness emerged within the very same abstractions—the techniques of infinite limits and infinite summation (which being algebraic were unthinkable within Greek mathematics)—that facilitated Galileo's overthrow of the Aristotelian account of motion and made it possible to defuse and bypass rather than refute the paradoxical conclusions that issued from Zeno's arguments. After all, if one can add, coherently and without contradiction, a completed infinity of time or space intervals to obtain a finite sum—for example, obtaining 2 as the sum of $1 + \frac{1}{2} + \frac{1}{4} + \frac{1}{8} + \ldots$ —there is, it would seem, no paradox extractable from an account of motion and subdivision resting on the necessity of such completed infinitary activities. But this was not to be: both Newton and Leibniz, using the language of physical movement to articulate the limit processes of the differential and integral calculus (points were spoken of as being "brought infinitely close together"), relied on the concept of an *infinitesimal* quantity, that is, a quantity smaller than any positive quantity and yet larger than zero. But, despite their centrality to the calculus, such infinitesimals (which, significantly, were completed infinities) proved to be confusing and frequently contradictory. At the beginning of the nineteenth century the attempt by Cauchy and others to develop a differential calculus for complex-valued functions (where the appeal to physical process and geometrically self-evident properties of diagrams was no longer possible) led to a rewriting of the discipline in terms of a finitely presented formalism—one based on a potential and not an actual infinitary process—that sought to exclude all mention of infinitesimal magnitudes and the physicalist language used to describe them. However, within this rewriting the continuum of points on a line and the field of real numbers representing them was itself conceived—unproblematically—as an actual, completed infinity.

The third moment, whose orbit still captures present-day thought, has its origins in the movement of criticism and foundational crisis initiated in the last quarter of the nineteenth century. Specifically, these origins lie in the rethinking of the numerical basis of the geometrical continuum by Richard Dedekind, in the regrounding of the whole numbers within pure logic by Gottlob Frege and within an axiomatic system by Giuseppe Peano, and, more spectacularly, in Georg Cantor's theory of aggregates which placed actual infinities—an actually infinite hierarchy of actual infinities—onto the mathematical scene as legitimate objects of mathematical investigation. The resulting radical reformulations of what was meant by a mathematical "object," by geometrical "curve," by the "dimension" of a space, by the "definition" of an entity, and by "mathematical reasoning," as well as attempts to deal with the various paradoxes and antinomies involving self-reference and infinite aggregates, brought the axiomatic method to the center of the mathematical stage and initiated the study of formal systems, metamathematics, mathematical logic, and computability. In so doing, it not only eventually produced the means of legitimating the concept of infinitesimals but, more importantly, resulted in the deepest, most searching, and most comprehensive reexamination of number, the geometrical continuum, and infinitary reasoning the subject has yet witnessed.

What, then, is the current status of the infinite in mathematics? What is its range and domain of meanings? What are the laws and possibilities of its signifiers? How *are* we, according to present-day mathematicians, to think and write it, either in its potential or its actual existence?

As observed, mathematics is dominated by Platonist orthodoxy according to which mathematical objects are metaphysically real, "out there," independently of and somehow prior to human cognition and discourse: by studying these objects that are given and existent before us, we discover or discern—but in no way shape, let alone create— what is "true" about them. The orthodox response, then, to inquiries about the status and nature of the infinite would be in predictably direct, no-nonsense realist terms. Infinity exists and is actual; the progression of integers is an obvious and undeniable example of an infinite set, the truths about which can be and have been extensively described and investigated (along with every other mathematical object) in a suitably formal, axiomatized version of the language of sets or aggregates employed by Cantor. Moreover, from the completed infinity of the integers comes the larger infinity of real numbers, as well as all the other infinitely many infinite sets, processes, and constructs required by contemporary mathematics. For Platonists, then, the modern crisis of infinity is essentially over, all doubts, circularities, and inconsistencies having been stilled and in effect swept away by a formal apparatus of definitions and axioms: at the third attempt, mathematics has finally settled the question of what the infinite is and how we are to think it.[14]

But this settlement is repudiated by constructivist mathematicians, for whom only the potential and not any actual version of infinity has any meaning. Before examining this we should note that, despite manifold divergencies over proof, logic, and ontological presuppositions, the division between Platonists and constructivists in relation to infinity might not be the total and unbridgeable methodological gulf it appears to be. Indeed, such a denial of Aristotle's fundamental opposition of the actual and potential infinite was Cantor's contention:

> There is no doubt that we cannot do without *variable* quantities in the sense of the potential infinite: and from this can be demonstrated the necessity of the actual-infinite. In order for there to be a variable quantity in some mathematical study, the 'domain' of its variability must strictly be known beforehand through a definition. However, this domain cannot itself be something variable. . . . Thus each potential infinite, if it is rigorously applicable mathematically, presupposes an actual infinite.[15]

Of course, given Cantor's mission of naturalizing the actual infinite and legitimating it as an object of mathematical thought, such a self-validating appeal to the prior acceptance of the domain of a variable—the progression of natural numbers considered as a unity, no less—is hardly unexpected. Interestingly, Blaise Pascal, writing at the time when the algebraic variable had just been introduced into mathematics, advanced a parallel form of argument but with the opposite rhetorical intent, concluding that the potential infinite, no less than the actual, belonged to the realm of theology and faith and not that of mathematics, which of necessity occurs within the secular boundaries of finite human reason.

But arguments like Cantor's or, were it to be taken seriously, Pascal's make little impact on constructivists' understanding of mathematical activity. So that against the Platonist settlement there is the constructivist dissent, the refusal of all the completed, actual infinities that appear so unproblematic, natural, axiomatically guaranteed, and objective to the realist gaze. For constructivist mathematicians, committed to articulating every mathematical procedure as a completed process that could, in fact or in principle, be carried out (according to some suitable regulatory scheme determining what it would mean to "carry out" a construction), only the potential infinity, the infinity of coming—endlessly to be sure—into being, can be cognized as meaningful and mathematically interpretable. That is, only the progression 0, 1, 2, 3, . . . of whole numbers escapes the wholesale extirpation of "infinities" from mathematics and survives as the one "natural" (potentially) infinite object. Accordingly, all constructivist programs and orientations for reconstructing or reworking mathematics, from at least Kronecker's explicit recognition of the fact onwards, base their strategies on the primality—presuppositional, ontological, and conceptual—of the integers on which the rest of mathematics is to be constructed. Thus one finds Kronecker's founding dictum of God's authorship of the whole numbers, inflected with a computational reading of "construction" but otherwise unchanged, a century later in Everett Bishop's programmatic declaration: "The primary concern of mathematics is number, and this means the positive integers. . . . Everything attaches itself to number, and every mathematical statement ultimately expresses the fact that if we perform certain computations within the set of positive integers, we shall obtain certain results" (1967: 2–3).

Such an understanding of the integers—that they are before us, given, "natural," elemental, forming the basis for the rest of mathematics and that (though requiring an axiomatic description to fix what we take to be their formal properties) they are in no need of a "basis" or, more to the point, any analysis of how they come to be—is the common property not only of the various constructivist approaches to mathematics but *a fortiori* of conventional Platonism too. The difference between the two is, for our purposes, merely a matter in the end of emphasis, cognitive style, and strategy for the subsequent development of mathematics. That is not to say, however, that constructivism—defined, in opposition to Platonism, precisely by its refusal to ignore questions of the manufacture and genesis of mathematical objects—is not more irresponsible for failing to question the genesis of the integers themselves. Nor is it to urge any kind of psychologism against constructivism: by "genesis" here is meant *linguistic production*, not psychological recreation of the integers, not the mental—introspectively validated—activity in the mind of some suitably idealized individual, but their coming into being as both source and consequence of mathematical discourse. What is required then is not, therefore, Brouwer's private cognition of the integers within the pure inner intuition of time as this is available to the transcendental Kantian ego,[16] but more radically an account of how, in doing mathematics, we can be seen—publicly and intersubjectively—to construct the integers in a way that makes them available to us all. One wants, in a certain sense, for constructivism to be constructivist about its own fundamental objects. Just such a demand was put to intuitionism by the logician Paul Bernays:

> From the integers k,l one passes immediately to k^l; this process leads in a few steps to numbers which are far larger than any occurring in experience, e.g., 67^{257}. Intuitionism, like ordinary mathematics, claims that this number can be represented by an arabic numeral. Could one not press further the criticism which intuitionism makes of existential assertions and raise the question: What does it mean to claim the existence of an arabic numeral for the foregoing number, since in practice we are not in a position to obtain it? Brouwer appeals to intuition but one can doubt that the evidence for it really is intuitive. Isn't this rather an application of the general method of analogy, consisting in extending to inaccessible numbers the relations which we can concretely verify for accessible numbers? (quoted in Parikh 1971: 494)

To pursue Bernays's request for a constructivism that would be constructive about its own ontology and not only about its reasoning, one would have to enlarge on the opposition that he makes appeal to here between actually producing a numeral "in practice" and being able in some general sense to produce it, as one says, "in principle." Now the relation and difference between performing actions (constructing, perceiving, inscribing, deducing) in "fact," in "practice," in "reality" and performing them in "principle," in "theory" is notoriously difficult. The difficulty, at least in the case of mathematics, turns on what it means to ask for and to depend on some consistent and workable difference between "large" and "small" finite magnitudes.

Not being able to answer this question seems to have stood as an insuperable obstacle in the way of giving any coherent or rigorous "activity-based" account of mathematics. Witness David Hilbert's failure within his metamathematical program to explicate the concept of a concretely presentable "finitary" proof, as opposed to one abstractly available within the classical infinitistic framework he was attempting to study. Or witness Wittgenstein's manifold and never-resolved problems in dealing with the concept of a visually graspable "surveyable" proof—an idea he was impelled to invoke as the necessary ground or principle for his calculational account of what we do when we understand and reason mathematically. Or, nearer the present point, consider the lack of any characterization by intuitionists, when they talk of proofs, of what it would mean to be able *in principle* to "perform a construction in the mind." And witness (though the relevance to the present discussion is less apparent and will be justified presently) the unresolved difficulties of the so-called sorites paradoxes associated with arbitrarily "large" chains of syllogisms. Of course, whether the present attempt to give an activity-based—that is, a language- and discourse-determined account of mathematics in which the activity in question is writing—will fare any better on these various scores than the efforts of Brouwer, Hilbert, or Wittgenstein, remains to be seen. More then of this later.

For the moment let us back off from any engagement with the manifestly difficult and troublesome opposition of practice/principle and return again to the question of how, according to present-day mathematics, we are supposed to think the infinite. The answers we are confronted with carry an inescapably complacent finality. Neither the Platonist confidence in a formal axiomatics that reproduces an already given and unproblematic infinite set of "natural" numbers as a posit and then identifies the members of this set as certain kinds of sets produced iteratively from the empty set nor the constructivist acceptance of this already-givenness as a basic, unexamined constituent of mathematical or premathematical thought available as a universal intuition seems to lead to anything beyond certain self-validations.[17]

But, as we shall see, it is possible to refuse the complacent finality of these positions and to ask questions about the nature of the "natural" which open up a very different picture of the whole numbers. This opening allows one to perceive behind the undoubtedly profound results about formal systems and the brilliantly elaborated technical achievements of twentieth-century mathematical logic and metamathematics a sense that, on a level more radical than that of formal syntax and axiomatics, the question of number—what it is, can be, and should mean—has been avoided; that the real issue of what we do when we bring numbers into being, and hence what we should or might mean by the mathematical infinite, has been passed over in silence; that on examination the current answers—classical or constructivist—amount to no more than a legitimation of business-as-usual as far as infinity, interpreted as endlessness, is concerned; that neither the constructivist dissent nor the massively defended Platonist orthodoxy ever moves outside the grasp of a metaphysical belief just as mysterious—but more hidden and subtle—as any they claim to have settled.

There are, to put the matter in the most direct and immediate terms, two layers of metaphysics that have to be negotiated here. The outer layer is essentially that signaled in the standard philosophical recognition of Platonism as a *metaphysical* realism. The metaphysics for us stems from its invocation of objects (in the present case, numbers) independent of discourse; objects that are outside any considerations of space, time, energy, materiality, entropy; objects which though non- or extra-human are such that, somehow, they allow mathematicians to discover and prove timeless truths concerning them—in a word, the understanding of numbers as heavenly, transcendental objects. Now, as observed already, much philosophical energy, from serious and elaborate critiques to subversive ridicule, has been spent attempting to overthrow the Platonist vision—with remarkably little effect on the beliefs of mathematicians. I have no wish to add to what is undoubtedly a fruitless enterprise.

Indeed, within the present essay there is no point for me to do so, since behind these critiques, untouched by them, lies in an inner, perhaps more fundamental layer of metaphysics wrapped even more tightly round the numbers. This concerns, not their status as heavenly Platonic objects, but the very idea of their *endlessness*, the idea of them as potential objects in an as yet unactualized but always and forever actualizable progression, waiting there in all their unlimited presence ahead of any kind of historical and symbolic engendering. For it is in this sense that their transcendental presence is as deeply embedded in all the available forms of constructivism as it is in Platonism. If the actual Platonic infinity seems to invoke some version of the Greek-Hebraic divinity, then this inner infinity of potential endlessness has to posit, as creator of the whole numbers, that divinity's—what we shall later see to be—no-less-transcendental angel. But "angel" puts too a fine a point on it: we will not be overly concerned to distinguish between the metaphysics of angels and a/the deity—between, that is, the being in question and the various signifieds of the term "God." After all, Kronecker was quite happy to name the creator of the whole numbers "God," and his constructivist heirs, who are the chief contemporary protagonists of the Platonists, though they perhaps prefer "natural" to "God-given," do not appear overly concerned to repudiate or disown Him.

But what if no deity created the numbers, what if they are not the work of a pure angelic being outside history, but rather the production of secular, time-bound, and empirically tainted culture? What would follow from a conception of the whole numbers as *humanly constructed*? Would we not at the beginning have to repudiate their closure in the past—God's creation of them in their entirety—and replace it with an openness framed in terms of the discursively manufactured present-future: numbers as things which exist or can be imagined to exist, which are humanly constructed and constructible, through some ongoing *symbolic process*?

But with this repudiation of their already-given entirety would not the "naturality," the much-prized "objectivity," the self-evident timelessness of the numbers immediately evaporate into the contingencies of culture, history, and the variability of human practice, leaving in their place a misplaced perception, if not a parody, of the objects in question? Are we not, as soon as we question their "naturality," rejecting a transcendentally guaranteed objectivity that seems to watch so assuringly over our experience of the numbers in favor of a temporalized rewriting of them that looks to be inseparable from the subjectivism and relativism so deliberately and successfully absent from mathematics? But might not this very opposition, presented as a choice of alternatives—absolute, natural, objective or contingent, subjective, relative—be a false standoff, the familiar raising of the relativist bogey by an objectivism under threat?[18] And if that is the case, might there not be a way through, or rather beyond, the opposition, a characterization of numbers which honored the sense of their felt objectivity without giving up on the insistence of their being brought into being by a process of human signifying activity?

One response to this question would be to examine the culturally determined content of "objectivity" itself, in particular, to try to show how the objective/subjective split in the case of mathematics is, no less than in other forms of knowledge, socially constructed. Some such approach, pursued from a variety of perspectives in the case of scientific knowledge over the past three decades, is essential in the case of mathematics if one is to overcome the sterile impasse of "naturality" that the Platonists and their constructivist opponents are content to operate with. Now, in a sense, the success of a broad and suitably formulated program of the sociology of scientific knowledge and its extension to the case of logic and mathematics is already an assumption in the present account. After all, to insist on thinking in terms of signs and discourse is to have accepted the social, cultural, intersubjective nature of mathematical reality as the basis for its description. Or, putting it less definitively, what is offered here is a specific—linguistic and semiotic—consequence of such a program that attempts to engage with the constructivity of mathematics in terms of the subject's characterization of itself as a symbolic activity.[19]

Thus, one must begin to try to answer the question: where do the whole numbers come from? And where can they come from, what other place is there to start with, but *counting*? The numbers arise—historically, psychogenetically, instrumentally, socially—in relation to acts of counting; they are seen to repeatedly issue from and accompany counting; and they are, as far as is possible to tell, incomprehensible in the absence of a complete familiarity with the particular repetitive signifying activity that we all seem to invoke by that term.

Counting presents itself as prototypical of the very business of sign creation itself. We count by repeatedly enacting the elemental process of creating identity by nullifying difference, repeatedly affixing the *same* sign "1" to individuated "things"—objects, entities—that are manifestly not the same qua individuals in the world-before-counting from which they have been taken. To talk of signs being, or being intended to mean, the "same" is to recognize their temporal existence, the maintenance of an identical signified through their history. The sign "the" in a text written a century ago, for example, is taken as the same—signifying the definite article in English—as the "the" whose tokens appear in the present sentence. Some mathematical signs—notably 1, 2, . . . —exhibit what is taken to be sameness-of-signified across large cultural milieus as well as historical periods. And, indeed, the belief in the absolute objectivity of such signs—the *total exteriority and independence of their signifieds to history and culture*—appears to constructivists and Platonists as self-evident and unproblematic.

Counting, then, taken as a mathematical ur-cognition, as the pure and distilled mode of the production of identity and sameness, is the site at which this belief will be most intense and resistant to any questioning. Now, as the nullification of difference through the assigning of "1," counting is inseparable from a sequence of ideal marks 1, 11, 111, 1111, 11111, and so on, which are the signifiers of what we might for the moment call "protonumbers." Having nullified difference to produce identity, the results of this procedure allow us to put into operation the reverse facility: we can, then, through the *introduction of new signifiers*, syntax-governed systems of names and notations, make distinctions and differences—odd, even, not-divisible-by-3, divisible-by-5, a perfect square, and so on—between formally distinct but qualitatively alike protonumbers such as 11111111111 and 1111111111 to arrive at familiar numeral expressions such as 11 and 10.

Clearly, the question of the infinity, the endlessness, of the whole numbers—or, in terms of a specific ideogram,[20] the meaning and purport of ". . ." in the expression 1, 2, 3, . . . —is transferred to the question of endless counting. Now counting, whatever else it might be, presents itself as a human signifying activity. So the question arises: how are we—through what discursive comparisons, appeals to examples, principles of cognition, modes of sign use, means of persuasion—to imagine an intersubjective business of repeating the same activity without end, of iterating the production and assignment of a signifier *forever*? Or, to put the question in the semiotic framework it demands and in the format we shall later use: *who* is, or is imagined to be, doing the counting? What sort of ideal agency or manner of semiotic being has to be invoked to discharge this endless affixing of identical signifiers? Who is the one-who-counts and, it seems, who-goes-on-counting (but this is the very question guiding our whole inquiry), without end?

The issues behind these questions will occupy us as soon as we start trying to spell out what it means and what it entails, as a business of interpreting and writing signs, to count and do mathematics. And, to make the status of such a business intelligible, we shall need to construct a systematic reading of mathematical signs, one rigorous enough to make the appropriate distinctions between various kinds of mathematical agency and sufficiently external to the way whole numbers are habitually cognized to resist the view of them as "natural."

Now, to cognize the whole numbers as given-before-us, with their endless continuation intrinsic to their "nature" and, as such, inseparable from their mode of being, is to operate according to what we might call the *ad infinitum* principle—the principle of always-one-more-time which allows the ability to ask one more question, carry out one more stage of a procedure, iterate an operation once more, take another single step forward, count yet again an item. The mathematized version of the ad infinitum principle (in formal arithmetic, the axiom that every number has a successor—for any number x there exists a number y such that $y = x + 1$) is the axiomatic source of the endlessness of the numbers: without it, it is impossible to preclude finite realizations of any intended axiomatization of arithmetic; with it, all models are infinite.

How could one possibly put such a principle into question? On what basis could one doubt the capacity to do something, identical to that which one has already done, another time—particularly if the action is so uncomplicated and straightforward as the unambiguous adding of a unit to an already formed number?

An immediate and obvious rejection of the ad infinitum principle—counting's equivalent of sudden death—is outright denial: simply call a halt to counting and declare by fiat that there is a largest number beyond which it is impossible to go, a number m for which the expression $m + 1$ is undefined. Such a move, not enacted but recurrently suggested as a kind of fantasy of finitism, is normally given plausibility by appealing, for example, to the supposed "finitude" of the number of individual particles in the universe and hence to the finitude of any collection of individualizable tokens available to instantiate the signifiers identified above as protonumbers. In a more agency-centered version of this, one might start not from the supposed finitude of the universe—whatever that means—but from some assumed set of limitations of the one doing the counting. Thus, one might disallow a counting process that was not performable within a given level of resource—for example, within the time-span of an individual life or by using only so much paper or time or storage facility.

But all such moves of constraint are, unless framed with respect to some *internal* aspect of counting, unacceptably arbitrary. Why one lifetime and not a hundred? Why this and not that level of resource? More seriously, they miss the point of setting out to question the ad infinitum principle. By being parasitic on the prior endlessness of the numbers in order to articulate themselves as coherent schemes, they appear as no more than artificial truncations, reduced models of an already existent infinite plenitude unable, therefore, to offer an independent critique of this plenum. Whatever their merits (for example, as projects for a finitized mathematics),[21] their account of counting reduces ultimately to a kind of Columbus phenomenon: counting proceeds homogeneously and regularly in identical repeated steps—exactly as if one were counting according to the ad infinitum principle—except that, abruptly and without any change in how one has been going, one falls off the edge.

Evidently, if one is to hold up the principle of always-one-more-time to scrutiny, it is not through attempts to curtail its scope by the introduction of what appears inevitably as an artificial cut-off point in an already given infinite progression of integers: such an external truncation merely begs the question of infinite prolongation. One needs, more radically, to put into question the supposed *necessity* or rational undeniability of the principle. But is it actually possible to do this, to deny the principle in a way that avoids collapsing back into the arbitrariness of an external rejection of it? According to contemporary mathematical wisdom—Platonist and constructivist—the answer would seem to be no. And even outside these fixed positions the answer seems just as strongly to be negative. Thus, the philosopher Hilary Putnam, highly critical of Platonism and by no means an apologist for constructivist accounts of mathematics, although he is prepared to exclude all manner of intuitive faculties and so-called rational principles from the canon of mathematical validity, draws the line at denying the possibility of always-one-more-time: "I do not doubt that *some* mathematical axioms are built into our notion of rationality ('every number has a successor')" (1983: 10). For Putnam, then (and one suspects for most other philosophers of mathematics), attempting to deny the ad infinitum principle in arithmetic would be to go— quickly and with little hope of retrieval—against reason.

But, as we shall see, this is not so. It is indeed possible to call the principle into question, to put forward a scheme of counting and hence of constructing—significatively creating—the numbers and interpreting them arithmetically which denies the necessity of one-more-time, yet does not fall into unreason and irrationality. In order to do this, what has to be subverted is not any built-in "rationality"— indeed, the critique of endlessness to be offered here never moves outside a quite narrowly drawn conception of the rational—but a certain understanding of pure *iteration*. One needs to see how and for what reasons one can refuse the idea of perfect repeatability; how one can justify an alternative to the idea of an unqualified, decontextualized, exception-free, universal *identity* between real or imagined human actions—an alternative and refusal that would operate no matter how idealized or fictionalized such actions are taken to be, provided only that they are accessible, in principle, to human practice.

Of course, as indicated above, much can and does turn on how we interpret what it means not to do something *in practice* but to be, nevertheless, able to do that thing (or in the context of numbers able to imagine doing it) *in principle*. One version of the phenomenon, focused on logic, occurs in acute form when the repeated action is that of drawing an inference. And when the content of the inference rests on a simple act of removal, it appears as the problem or paradox of sorites (the "heap"). The standard formulation of this runs as follows. Subtracting a single grain from a heap of grains leaves in place a heap—such is the meaning of "heap"; each further subtraction will likewise leave a heap. Thus either a qualitative change from heap to non-heap occurs at some particular point, contrary to supposition, or one is left with the absurd result that after all the grains have been subtracted there will still remain a heap. Another version appears as the argument that because not being bald is unaffected by the removal of an individual hair, one cannot be rendered bald by the repeated subtraction of single hairs. Though the paradox—a mismatch between what syllogistic logic predicts as the outcome and what is manifestly the case about some aspect of physical reality—is evident enough in these and numerous other examples, there seems no accepted explanation of the paradox's source.

One major approach, mindful of the absence of sorites phenomena from mathematics, sees the problem in the inherent vagueness of predicates like "heap," "bald," and so on, and urges the need to develop a logic and a notion of truth—different in some way from that operating for the totally precise and exact predicates of mathematics—to deal with them.[22] But, whatever the merits of this analysis, it doesn't focus on what is in the present context the key feature of the phenomenon, namely the role of iteration *as such*. The paradoxical difference between logic and reality produced by a sorites argument is plainly a function of repetition: each time one repeats the initial syllogism (by making the conclusion of one stage serve as the premise of the next), the divergence between prediction and actuality widens. This suggests recognizing a change in the status of the action (removing a grain) with respect to the predicate (a heap) each time it is in fact carried out—in *fact*, as a result of actually performing it, and not in *principle*, not in the prior theory of a logic that can comprehend no possible difference between x and $x + 1$ successive applications of its rules. Such a recognition opens up a gap inside the logic of statically conceived, context-free "equality," since decrements and increments are seen to be not equal in their cumulative effects even if they are "equal" as signs designating the "same" activity.

The relevance of the sorites phenomenon to the present account of mathematical infinity lies precisely in the nature of this gap. The challenge to the *ad infinitum* principle arises out of the possibility that the repetition of making a mark is not a homogeneously unchanging business. If it were, there might indeed be no rational, internally constituted obstruction to always doing it one more time. But if, on the contrary, the operation of adding a mark—the identically "same" operation applied to the "same" mark—were to be understood as cumulatively productive of a *difference* not itself definable on the level of these identities—a difference that made a difference,[23] one capable of introducing qualitative changes (ultimately bringing the entire process to a "close")—then a radically new interpretation of counting and construction of the whole numbers presents itself. And, associated with this interpretation, there emerges the possibility for the formulation of *non-Euclidean* arithmetics.

The idea of such arithmetics seems strange enough to immediately raise a swarm of questions and possible objections. What could their internal structure be to justify the claim that they are "non-Euclidean"? How is one to articulate the relationship between the integers that form their object and the ordinary, "natural" numbers of the current—Euclideanly conceived—classical arithmetic of identical, homogeneous increments? What would be their comparative instrumental value as an alternative mode of measuring and enumerating physical reality? All these will occupy us later, after the semiotic basis for the construction of an alternative arithmetic has been developed. At this point one can do no more than make some anticipatory gestures in the context of the difficulties and issues crystallized in the sorites phenomenon. Thus, as soon as a given addition of "1" is contextualized—that is, seen not as identical to any other such addition, wherever and however it occurs, but dependent on the cumulative effect of the sequence of additions needed to get to the point where the given one becomes possible—the nature of counting is drastically altered. Iteration is seen as subject to what might be called a cumulative resistance whose source and effects arise as a phenomenon purely internal to the process of counting: the production of integers via repeated addition of "1" becomes increasingly difficult, so that one reaches a region where their further production through such repetition becomes impossible. This will imply that the expression "+1" can have different possible significations, depending on its context, so that, for example, its effect in the notation $9 + 1$ might, under certain stated arithmetical conditions, coincide with its effects in $99 + 1$, $9,999 + 1$, $99,999,999 + 1$, and so on, *but cannot be assumed or guaranteed to do so*, in any universal, unconditioned, and "total" sense.

The contrast between the global properties of the numbers produced under this new, contextualized interpretation of the successor function and their classic counterparts runs parallel to the opposition between real heaps, which collapse into piles or spillages before disappearing, and the fictive, never-shrinking models of them invoked by classically sanctioned iteration of syllogistic logic. Like actual heads of hair that can change shape, develop bald patches and pass through innumerable stages under the repeated loss of one hair at a time, the non-Euclidean numbers exhibit qualitative changes, identifiable and definite shifts in their mathematical properties, under the repeated impact of adding "1." What can be said and proved, in terms of the arithmetical operations of addition, multiplication, exponentiation, and so on, about "large" numbers will increasingly diverge from the familiar arithmetical laws which we recognize and agree to be the case about "small" numbers. Within this perspective the denial of the ad infinitum principle manifests itself as a feature built into the very business of iteration and not something that happens at some particular—and, more to the point, externally specifiable—stage. One cannot expect, in other words, to point to individual numbers at which shifts in the structure of numbers occur any more than one can single out which individual grain's or hair's subtraction "causes" a qualitative change. It is always a question (for both heaps and numbers) of zones and regions, of the boundaries of relations between global and local characteristics, of the emergence of novelty at a higher level than that at which particular numbers are cognized, of the increasing presence of indeterminacy and open-endedness as one iterates again and again and . . .

But granting all this for the present (granting the possibility of systematically denying the subordination of counting to a pure un-changing Euclidean iteration, granting the rational viability of a scheme capable of rewriting the ad infinitum principle, and on the basis of which thereby developing some kind of alternative arithmetic), where does this leave the *infinite*? Establishing the possibility of denial here is surely not to be taken as a repudiation of the classically conceived Euclidean infinite as such, any more than creating and exhibiting the coherence of non-Euclidean geometries does away with the ideational content of the standard Euclidean formulation or the purely internal, mathematical validity of its propositions. That said, is there in this denial of the axioms of Euclidean arithmetic any real challenge to the *idea* of infinity? Do we have here merely a certain kind—more or less interesting, more or less cogent, plausible, fruitful, and so on—of empirically based "finitism" that, for all its radical implications about iteration and perfect identity, misses the point—namely, that infinity as a *mathematical concept* has to be seen as referring to an unchang-ing, *ideal*, nonempirical process which—by definition—is outside any effects of dissipation, degradation, or entropic loss of identity?

Such, or something very like it, might be the response by con-temporary mathematicians—Platonist, constructivist, or any other persuasion—to the scheme for treating infinity being outlined here. Understanding what is inadequate about this response and why in the end it is circular and self-validating is complicated by several issues. First is the question-begging impasse ineradicably built into any un-critical acceptance of the term "finitism." Then there is the equivoca-tion that is integral to the conception of "axioms" inherited from Euclid, as to whether they are free-standing mathematical postulates or so-called truths about "the world." And finally, there is the unexamined nature of the *ideal* and how one is supposed to arrive at what is and isn't to be included within the limits of a mathematically acceptable process of idealization.

The term "finitism," like "atheism," rests on the denial of a prior absolute; in mathematics the absolute is the already existent plenum—God-given, no less—of the unending progression of natural numbers. Finitism as a discursive ascription, then, is an essentially posterior and privative move connoting a lack, a falling short from some prior, fully existent object, a refusal of sense—expressed within the language of what is being denied—to an already intelligible state of affairs. So that if this object and state of affairs is itself at issue, if the prior existence and prior intelligibility of the infinite plenum is precisely the problem, then the question of how, as embodied beings immersed in material signifiers, we are to think the infinite is nullified at the very beginning. What appears as a neutrally descriptive term exterior to the discussion can never be other than a self-enhancing rhetorical maneuver by an infinitism inuring itself against any criticism of its status. In a sense it is "infinitism" that should be under question with respect to a prior finitude; but this reversal in the burden of proof, though correct on the level of pure logical ordering, is not feasible, since what constitutes a normal interpretation of "finite" is already folded in to the term "infinite." A better term than "finitism," though we shall not use it, might be "actualism," since the structure of the argument here proceeds from the actuality of human counting embedded in speech and the written materiality of mathematical discourse. Only after that—on its basis and in relation to the demands and signifying possibilities opened up by it—can the finity/infinity question of what is to be understood by the ideogram ". . ." come into being.

Consider next the Euclidean equivocation. Since its axiomatic formulation in classical times, geometry has had a dual aspect, as both deductive system which works out the logical consequences of a chosen system of axioms and as a proposed or postulated description of the nature and structure of (our experience of) space. From Euclid until the nineteenth century these two interpretations—logical axioms and descriptive postulates—were fused, it being held that the axioms or postulates set down by Euclid were true statements (somehow necessarily so) and, since logic preserved truth, that the derived theorems were likewise truths about "space." The advent of non-Euclidean geometries forced this equivocation between logical posits (adopted, changeable postulates) and descriptive claims (necessary truths) into the open and instituted the necessity for a methodological separation. Any system of geometrical axioms became simply a theory or model, more or less accurate, adequate, or useful, for interpreting an observed, empirically given space which, for all one knows or can know, may not be capturable or adequately cognizable within any axiomatic system whatsoever.

The obvious question arises: Is there not a parallel form of equivocation at work in the way the classical axioms of arithmetic (notably, in the present case, the stipulation that every number has a successor) are cognized? It seems difficult to escape the sense that these axioms too are normally understood as both logical posits and as (somehow necessary) truths about their subject matter, as both *a priori* and synthetic in Kant's vocabulary, as descriptive of number but also as being "built into our notion of rationality," as Putnam has it (1983: 10). If this is so, then the possibility of a separation between logical posits and descriptive claims with respect to time and its theorization through arithmetic (reflecting the separation between space and its geometrical theorization) presents itself. Certainly, as soon as one recognizes that numbers are unintelligible except in relation to counting and that counting, however idealized, is a *temporal* process, there is a prima facie case for the proposition that arithmetic is no less a model for the structure of time as it governs actual or actualizable process than geometry is a model for empirically presented extension in space.

If such a project of separating posits from "truths" were to be carried out, there would result a coherent and workable distinction between the current Euclidean arithmetic and a variety of non-Euclidean versions; the latter's axioms would appear not as necessary truths but as theorizations of a process—appropriately idealized—of iteration. Of course, there would still remain for arithmetic the issue alluded to earlier for geometry, namely the relation between the Euclidean and non-Euclidean versions. Such a relation would range from the pole of instrumentality (no arithmetic that failed to reflect the practical, elementary "truths" of ordinary, Euclidean arithmetic, such as $1 + x = x + 1$ for any number x, could be taken seriously as an explication of anything we could call "number") to the pole of abstract theorizing (in which the ideogram ". . ." of so-called endless continuation would be reinterpreted). We shall return to the problem of explicating these relations.

Finally, and most crucial of all, there is the question of idealization. If mathematical infinity is *essentially* an ideal entity—a transcendent phenomenon of pure iteration totally divorced from all considerations of spatio-temporality, untouched by the limits of the empirical and the material that govern all concrete instantiations of number and outside the problems of *actual* repetition (such as occur for example in the sorites paradoxes)—then indeed, whatever its interest, the account being constructed here fails to impinge significantly on it. But what, if anything, is the "essence" of the ideal? How ideal can an ideal—that is to say an idealized—entity be? Does idealization, which invariably connotes a passage to some limit, itself have a limit? Is it not possible for the *process* of idealization—the very business of abstracting all empirical, actual, real content from an object and cognizing the result—to violate the relevant integrity of the object, to replace it by a fictive surrogate that bears no relation to that object with respect to the purposes for which the idealization has been set in motion? One can certainly idealize iteration into "endlessness" and construct for the resulting ideal a transcendental realm—such is the achievement of classical infinitism—but what is the relation of this ideal to the matrix of embodied, unidealized repetition from which it is supposed to derive its sense?

To understand the way in which there could—must—be boundaries and limits, points of rupture marking regions of qualitative change inherent in the process of mathematically idealizing the "same," one has to return to the semiotic question of a sign-using

MATHEMATICAL INFINITY / FINITY

agency. If numbers result from counting, and if counting is to be cognized as an idealized repetition, then what sort of semiotic figure or being is to carry out this repetition? Who is the one-who-counts? Obviously, such a being will be imagined, will be a fictive proxy for the empirically constituted, corporeal mathematical subject who sits down to read, write, and count mathematical signs. The question is: what connection is there, in the case of "endless" counting—other than the declared intention that there be one—between this fictive being and the embodied mathematical subject it "infinitely" transcends?

. . . 3

EXPERIMENTAL THOUGHT

Still more astonishing is that world of rigorous fantasy we call mathematics.

· Gregory Bateson, *Steps to an Ecology of the Mind*

Since Galileo the thought experiment (*Gedankenexperiment*) has been a widely used element of scientific theorizing. Only fairly recently, however, have these scenarios and narratives of imagined activities—actions realizable in principle but difficult or unfeasible to execute in practice—begun to receive any serious attention. This is no doubt because, being used as devices of illumination, explanation, and persuasion, they have been seen as merely rhetorical in relation to so-called proper explanation and *real* experiments. The putting into question of this "merely" as part of an overall recognition of the role of persuasion in the constitution of scientific practice is an historiographical achievement of the last two decades. The historian Thomas Kuhn, for example, has directed attention to the constitutive function of thought experiments, observing how they have "more than once"—in the mid-seventeenth and early twentieth centuries—"played a critically important role in the development of physical science" (1977: 240).

From our earlier discussion, any characterization invoking a separation between actions performable "in principle" and those executable "in practice" should alert us to the possibility that certain deep-rooted difficulties are at work. In any event, Kuhn puts to the idea of thought experiments a series of questions:[24]

1) Since the situation imagined in a thought experiment clearly may not be arbitrary, to what conditions of verisimilitude is it subject? In what sense and to what extent must the situation be one that nature could present or has in fact presented?

2) Granting that every successful thought experiment embodies in its design some prior information about the world, that information is not itself at issue in the experiment. On the contrary, if

we have to do with a real thought experiment, the empirical data
upon which it rests must have been . . . well-known. . . . How,
then, relying exclusively on familiar data, can a thought experiment lead to new knowledge or to new understanding of nature?

3) What sort of new knowledge or understanding can be so
produced? What, if anything, can scientists hope to learn from
thought experiments? (1977: 241)

What I propose here is a semiotic model of mathematical activity
fabricated around the idea of a thought experiment. The model identifies mathematical reasoning in its entirety—proofs, justifications, validation, demonstrations, verifications—with the carrying out of chains
of imagined actions that detail the step-by-step realization of a certain
kind of symbolically instituted, mentally experienced narrative. Thus,
unlike physical science where thought experiments are contrasted
with real experiments and are seen as one ratiocinative and persuasional device among many, mathematics, as presented here, will be
singularly and exclusively founded on them. As a result, the problems
we encounter will not be identical to those behind Kuhn's questions
but rather transformed versions of them. Thus, in relation to the third
question, the issue of what can be learned from thought experiments
becomes altered, since by extending to the whole of mathematics it
asks for an account of mathematical thought itself. Moreover, the
term "knowledge" is, according to the perspective here, a misleading
description. With respect to Kuhn's second question, which is about
the context and use of prior knowledge, the issue becomes that of the
relation between the mathematical Code and natural language and, in
particular, the route by which mathematics achieves its instrumentality. And finally, in relation to the first question, the issue of verisimilitude will be here the crucial problem of idealization. More specifically,
in terms of the model we are about to present, the issue concerns
the criteria that it is claimed must be satisfied by the imagined simulacrum of the sign-using subject.

It was Peirce who seems to have first suggested that mathematical
reasoning was akin to the making of thought experiments, "abstractive observations," he called them; and Peirce's contemporary, the

physicist Ernst Mach, not only saw in them essential tools of physics which gave a kind of experimentation without apparatus but also, like Peirce, understood them as the basis for any kind of planning and forward-directed rational thought.[25] Obviously, the relation between real experiments and thought experiments will be significant here, and to this end it will be helpful, before introducing Peirce's characterization, to go back historically to the point when the category "experimental" was itself being legitimated as the practical basis and founding abstraction of empirical science.

In their study of the origins of English experimentalism, as this was instituted by Robert Boyle in the late seventeenth century, the historians Steve Shapin and Simon Schaffer set out to demonstrate that "the foundational item of experimental knowledge"—that is to say, the very category of scientific "fact,"—"and of what counted as properly grounded knowledge generally was an artifact of communication and whatever social forms were deemed necessary to sustain and enhance communication" (1985: 25). To this end they identify several technologies—"knowledge-producing tools"—employed by Boyle to establish the categorial identity and legitimacy of scientific "facts" and their separability from purely theoretical observations. The chief tool in the present context is what they call the "technology of virtual witnessing"—a rhetorical and iconographical apparatus, a carefully constructed way of writing and appealing to pictures and diagrams which, by causing "the production in the reader's mind of such an image of an experimental scene as obviates the necessity for either direct witness or replication" (1985: 60), allowed the "facts" to speak for themselves and be disseminated over the widest possible arena of belief. Behind the employment of this apparatus was Boyle's anxiety to distance his method of empirical verification—and with it the claim that such a method produced and dealt in the natural category of "facts"—from the kind of experiments involving only pure "ratiocination," unconnected to the rigors of actual experimentation, that he attributed to others, notably Pascal. Now Pascal was, after all, a mathematician and experimental ratiocination—carrying out thought experiments—is, I urge, *precisely* what mathematicians do.

In other words, the proposal here is that Boyle's rhetorically accomplished replacement of the actual by the virtual, the shift from
doing/watching to the imagining of doing/watching, the move from
real, witnessed, and executed experiments to their virtual, reproduced-
in-the-mind versions charted by Shapin and Schaffer points, contrary
to Boyle's efforts to separate them, to a certain commonality—a duplication of persuasive technique—between the establishment of empirical *factuality* and mathematical *certainty*. Each results from the use of
an elaborately designed apparatus able to mask its discursive and
rhetorical origins under the guise of a neutral and theoretically innocuous method for discovering naturally occurring "facts" or, in the
case of classically conceived mathematics, objective "truths" about
the so-called natural numbers. Identifying the thought experimental
mechanism for mathematics will provide, then, a fruitful way of explicating the persuasional basis for mathematical reasoning and hence
for understanding how mathematical logic is to be seen as inseparable
from a species of rhetoric.

To be sure, there is a fundamental difference between mathematics and physical science and between real experiments—however their
results are rhetorically processed and disseminated—and imagined
ones that needs to be understood, and we shall return to the question
of their relationship at the end of this chapter and, more schemati-
cally, throughout Chapter 4. One thing immediately clear, however,
is that what corresponds in mathematics to the empirical reality
engaged with by science—mathematics' "reality"—is already a symbolic domain, a vast field of purely fictive objects. One consequence
of this is that the very difference between direct manipulation of such
symbolic objects and virtual or thought experiments on them threatens to be lost in the space between imagining actions and imagining
imagining them. To avoid this possibility, one needs, as we shall now
see, to pay careful attention to the language of mathematical texts
and, in particular, to separate out the agencies they invoke in their
instructions to manipulate mathematical signs.[26]

As an ongoing cultural endeavor, mathematics works through a process of communication. Mathematicians circulate and exchange signs with each other, with themselves, with past and future members of their community in the course of responding to contemporary and historically posed questions, proving new theorems; expounding, elaborating, axiomatizing, explicating, modifying, restructuring, and re-cognizing existing mathematical objects; making conjectures; giving explanations, motivations, definitions; suggesting exceptions, puzzles, and counterexamples; and so on.

As part of the normal process of self-definition—the fixing of disciplinary boundaries, the legitimation of procedures which establish what is and isn't to count as their subject—contemporary mathematicians divide their activity (without theorizing the business or being very explicit about it) into two modes: the formal and the informal.

The *formal* or rigorous mode—privileged as mathematics "proper"— is made up of all those *written* texts (books, papers, research announcements, blackboard lectures) which are presented according to very precise, completely unambiguous rules,[27]—conventions and protocols controlling the permitted patterns of sign use with respect to the content of assertions and their logical validation, the status of conjectures and hypothesis, the giving of definitions, specifying of notations, exhibiting of objects, and so on. In the extreme form of laconism permitted by mathematics such texts consist of sequences of definition, theorem, proof, and nothing else. I shall call the sum total of all these resources—that is, the unified system of all such rules, conventions, protocols, and associated linguistic devices which sanction what is to be understood as a correct or acceptable use of signs by the mathematical community—the *Code*.

The *informal* or unrigorous mode makes up the mass of signifying and communicational activities that in practice accompany the first mode of presenting mathematics: drawing illustrative figures and diagrams; giving motivations; supplying cognate ideas; rendering intuitions, guiding principles, and underlying stories; suggesting applications; fixing the intended interpretations of formal and notational systems; making extra-mathematical connections—generally using natural language as well as a variety of explicational moves, tropes, and iconographical devices to convey all manner of mathematical sense, ranging from the level of technical content to the "point" of a proof or the "triviality" of a result or the "fruitfulness" of a counter-example. This heterogeneous and divergent collection of semiotic and discursive means I call the *metaCode*.

The image that seems to motivate this division would appear to be that of an inner core of rigorous, fully illuminated "real" mathematics surrounded by a penumbra of vague, not very well lit facilitating devices—stories, psychological hints, heuristics, figures of speech, motivations, gestures, pictures—whose function is to ease the difficult ascent into the space of pure reason. Such a function, though it would seem in fact always to be necessary, is held in principle to be eliminable, a matter in the end of mere affect and emotional resonance, essentially supplementary and epiphenomenal to the real business of doing mathematics that takes place in the Code and incapable of entering into or even leaving any impression on the content of what it facilitates. In effect, the Code—the site of pure writing—is seen to be precise, logical, objective, ordered, atemporal, rational, and un-changeable as against the metaCode—the place of speech and ges-ture—which is vague, permeated by illogic and subjectivity, intuitive, uncategorized, immersed in history and cultural mutability. I shall argue below that this understanding of the metaCode, as the dimness before the light, the practically necessary but theoretically dispensable stage on the route to unadulterated truth, relies on and fosters a deep-rooted misconception of how mathematical objects get signed into being, and that, on the contrary, Code and metaCode enter into each other, that all mathematical acts—however rigorous and formally self-sufficient they might appear—require, pass through, and are con-stituted by the prior presence and structure of the metaCode.[28]

Mathematical signs, formal and informal, are by definition inter-subjective: they circulate through, are sent by and addressed to sub-jects. In the case of the Code I shall call the common, idealized sign-using agency—the universal addressee assumed and put in place by the Code—the mathematical *Subject*; and, quite distinct from this, by the term *Person* I shall refer to the universal addressee/user required by the metaCode, the one who speaks through and by virtue of it and who has access to the "I" of natural language that permeates the metaCode. To understand the relation between these agencies as well as their joint contribution to the figure ordinarily and uncomplicatedly called the "mathematician," we need to look first at the grammar of mathematical texts. Even a cursory examination of an arbitrarily chosen item of mathematical communication will reveal two fundamental features of mathematical discourse: its organization as an exhortatory, command-giving formalism and its complete lack of any indexical terms.

First, exhortation. Our chosen item of communication will be an amalgam: syntactico-grammatical devices, special symbols, ideograms, notational constructs being used in what is evidently a systematic way, together with fragments of some natural language—English, Japanese, Hindi, or whatever—that will be saturated with imperatives dictating that all manner of orders, commands, and instructions, detailing particular operations, be carried out in relation to the relevant ideograms: "construct the closure of the field P"; "consider $1/x$ for all non-zero integer x"; "prove that $a = b$"; "exhibit all real roots of the polynomial f"; "drop a perpendicular from each vertex of the figure D to the opposite side"; "reverse all functors in the category C."

Despite the enormous variation—in conceptual type, scope, domain of applicability, reflexivity, constructiveness—between different mathematical imperatives, they can for the purposes at hand be assigned to one of two kinds, corresponding to a distinction drawn by linguists between the inclusive imperative ("Let's go!") which asks for an action including the speaker to be carried out and the exclusive imperative ("Go!") where only the hearer and not the speaker is involved. According to this, the inclusive imperative in mathematics would be made up of "(let us) consider/define/demonstrate/prove," and their synonyms, all of which ask that a common world be instituted or inhabited or be the focus of some shared concern. The exclusive imperatives—in effect the remaining mathematical commands— dictate that certain actions meaningful in these worlds be executed.

Thus, for example, the inclusive imperative "consider a category *C*" is an injunction to establish a shared domain of "categories"; it demands that its addressee introduce a certain standardly conceived network of basic signs, definitions, proofs, and examples. These will involve arrows, commuting diagrams, objects, morphisms, functors, natural transformations, and so on that together determine what it means to inhabit the world of categories and bring into play what is needed before any—exclusive—imperative concerning categories can be executed. By contrast, the imperative "integrate the function *f(x)*" is purely exclusive; it assumes that a shared world, that of the calculus, has already been instituted and asks for a specific action relevant to this world, namely the process of integration, to be carried out on the function *f(x)*.

To whom are these various imperatives addressed? Who obeys these injunctions and carries out the actions they specify? After all, unlike the situation in physical science where an addressee's activity can involve observations, experiments, and measurements of external reality, the only actions mathematicians qua mathematicians can normally be supposed capable of are those of thinking/imagining and reading/writing: inscriptions with specific interpretations and imagined content are manipulated according to conventions which both honor and determine this content. In the case of inclusive imperatives, which ask for certain structured sign configurations—proofs, definitions, notations—to be produced, written, and brought into appropriately intentioned play, the addressee is the Subject, the one who reads, writes, and uses these signs in conformity with the various rules, protocols, and conventions of the Code.

But what of exclusive imperatives: who is obeying these? To whom are these technical rather than frame-setting orders addressed? Clearly, it is not the embodied sign-reading Subject who can be asked to enumerate the rational numbers or invert an arbitrarily large matrix or execute the endless process of addition involved in integrating a function and so on. Rather, it is an idealized and truncated version, a model or simulacrum of the Subject, imagined into existence by the Subject in order to execute actions that go beyond—in a sense yet to be explicated—the physical and cognitive capacities of the Subject as an embodied being. I shall use the term *Agent* to denote this simulacrum who, by executing activities the Subject imagines for it, serves as the mathematical proxy for the Subject. One might say that the inclusive and exclusive imperatives correspond to the mutually collaborative use and the mechanical execution of sign operations respectively. The sense of this distinction will emerge when we come to examine the different roles played by the Subject and the Agent in the course of a mathematical proof.

Second, indexicality. An examination of the locutions permitted by the Code to figure in a mathematical text reveals the complete absence of any of the familiar indexical signs such as "I," "you," "here," "now," "this," and so on. At no point during a mathematical argument or in a definition or in relation to any mathematically specified action or sign operation that takes place within the Code is there any reference to the time or place, the cultural, social, or historical moment, the physical circumstances or psychological reality in which such signifying takes place. The Subject's immersion in publicly identifiable space and private durational time—necessary and unavoidable for any sign reading/writing being to function at all—is simply not registered or indeed registrable within the Code.

This lack of an indexical apparatus on the part of the Code—which not only seems to entail the exteriority and irrelevance of the Subject's spatio-temporal presence to mathematical content but also has the effect of severing the constitution of the Code from the context and circumstances of its practice—is by no means accidental. Over the course of the development of mathematics, particularly since the scientific revolution of the seventeenth century, mathematicians have made it their policy to exclude from mathematical language any mention of empirical reality: whole programs of "rigor" have been devoted to eliminating physicalist language (e.g., a curve as the path of a moving point) in favor of purely formal, internal characterizations (a curve as a certain kind of function from the continuum to itself). No doubt an external driving force behind the search for rigor was the need to deliver to science a formalism completely free of any subjectivist and physicalist presuppositions about a supposedly prior and unsuppositioned reality (an issue to which we shall return in Chapter 6).

But, whatever the motives and the nature of the historical processes embodying them, the end result of these programs has been to present a mathematical Subject who, though an embodied, imagining, and intentional sign-manipulating being, cannot articulate any of these features of itself through the signs made available to it by the Code. The Subject, in other words, has been denied the subjectivity available to the Person through the latter's access to the indexical "I" in the metaCode. And in this lies the principal operative difference between the Subject and the Person as semiotic agencies. This denial manufactures the Subject as an a-historical, a-cultural, a-social truncation of the Person, a nameless trans-individual operating under a universal psychology. Mathematical texts, then, present us with an I-less Subject, a figure who speaks about—asserts, proves, considers, defines—number, space, and time in the language of an unlocated, always-already-there present: a timeless voice from no one and from nowhere.

Were we to leave matters at this point, we would have arrived at a description of mathematical texts no more satisfactory in some ways than that given by conventional Platonism—a description which leaves as an unsolved mystery the problem of *access*: how do corporeal mathematicians have access to the disembodied, a-temporal realm of "truth"? How, in the present terms, is one to reconcile the demands of intentionality and embodiment inherent in the very idea of a Subject who reads/writes meaningful, material inscriptions with the manifestly unlocated, nonindexical textual presence of this Subject? Indeed, is the conception of a disembodied "I"—or, conversely, of an I-less corporeality—as a self-contained explanatory semiotic agency not in some way incoherent or perhaps contradictory? The answer is no, but it is not simple to explicate it. We have to elaborate the issue raised here of the Subject's supposed "self-containment"; and before we do this, there is the closely related but separately posable question of whether mathematical texts can *ever* eschew all reference to, or semiotic reliance on, physicality.

On this latter question observe that a sine qua non of all mathematical activity is the ability to initiate the process of counting in order to bring the numbers into being and, consequent on this, the ability to initiate and put into place a coordinate system with respect to some suitably cognized space. Both these initiating acts are signified within the Code by zero: the sign "0" being used to denote the starting point or origin of the sequence of numbers, and the same sign "0" or its higher dimensional analogues—(0,0) or (0,0,0) and so on—to denote the spatial origin of coordinates.

The notion of a starting point or "origin" is revealingly problematic, since it seems impossible to cognize it except as a process of recognizing it, of finding in it, if not the presence, then the trace of the presence of an indexicalized cognitive act. Such a recognition seems to lie behind Herman Weyl's curious and otherwise somewhat puzzling description of the mathematical origin as "the necessary residue of the extinction of the ego" (1949: 75). Such a previous act of cognition, outside the Code which circumscribes the Subject, could only have come from the metaCode and be the work of the Person. This would entail a dual role for zero, since it would both be a sign in the Code, where it would feature as a number among numbers, as in $1 - 1 = 0$, $3 + 0 = 3$ and so on, and also belong to the metaCode as the "origin," where it would be a metasign marking the role of the Person as the one who has already initiated the business of counting. Now the question of zero as an "origin," as a sign and metasign of initiation, gives rise to a complex and intriguing story which I have pursued elsewhere (1987). The point to be made here about "0" is that in practice its two senses—Coded number and metaCoded origin— are inextricable from each other: in the course of manipulating the number sign, the meta-sense is always present shadowing it, being part of another layer of meaning which adjoins and penetrates the formal layer available in the Code. This Code/metaCode interweaving, whereby a formal sign is doubled by an external, pre-image of it in the metaCode, though highly focused and singular in the case of "0", is in fact, as we shall see, the condition of all mathematical signs—not least the sign ". . ." whose critical unfolding is our goal here.

Mathematicians, we have agreed, can be said to act only insofar as they read/write and think/imagine. The picture presented so far of the second of these activities has a Subject imagining—as the result of responding to an inclusive imperative—a world into being and then assigning a proxy, an Agent, to perform various—imagined—actions in these worlds. Such a process is, of course, nothing other than a thought experiment.[29] Or, as Peirce calls it, an "abstractive observation" that we set up to answer hypothetical questions:

> It is a familiar experience to every human being to wish for something quite beyond his present means, and to follow that wish by a question, 'Should I wish for that thing just the same, if I had ample means to gratify it?' To answer that question, he searches his heart, and in so doing makes what I term an abstractive observation. He makes in his imagination a sort of skeleton diagram, or outline sketch of himself, considers what modifications the hypothetical state of things would require to be made in that picture, and then examines it, that is, observes what he has imagined, to see whether the same ardent desire is there to be discerned. By such a process, which is at bottom very much like mathematical reasoning, we can reach conclusions as to *what would be* true of signs in all cases. (Buchler 1940: 98)

I have introduced three agencies—Person, Subject, Agent—whose joint actions constitute the armature of any mathematical thought experiment. The role of Subject as imaginer, identified by Peirce as the "I" or "self," and the function of the Agent, the one who is imagined, the "skeleton diagram or outline sketch" of the self, as Peirce has it, are evident enough. In terms of signifying capacity, observe that the Agent, unlike the Subject, has no ability to imagine and can only respond to signs in their truncated, skeletonized form as signifiers devoid of intentioned meaning. In other words, the Agent is conceived as an automaton, a wholly mechanical and formal proxy for the Subject. But what of the Person? Peirce makes no mention here of a third agency. Ought he to have done? The answer hinges on the nature of logical conviction and persuasion operating here: who is being persuaded in a mathematical thought experiment and, before that, of *what* are they being persuaded?

As models of mathematical reasoning and of proof, thought experiments are addressed to *assertions* of mathematical content; that is, they are validations of claims that some particular situation "holds" or "is the case" or "is true." And what constitutes a claim is a syntactical construct built up from a small number of basic claims such as "$x = y$," "there exists x such that P," "all x have property Q" and so on. Thus the following are typical assertions: "$3 + 2 = 2 + 3$"; "for all integers n the equation $n^2 - 1 = (n - 1)(n + 1)$ holds"; "some topological spaces are not separable"; "the number 1,000,000,000,001 is prime." Such claims take the form of predictions; they tell what will happen if certain sign manipulations implicit in the linguistic form of the claim are carried out. Thus, the assertion "$2 + 3 = 3 + 2$" predicts that the result of concatenating 11 and 111 will be the same as concatenating 111 and 11; the assertion that the equation $n^2 - 1 = (n + 1)(n - 1)$ holds predicts that whatever value of n is used, the computations on either side of the "=" sign will give interchangeable results; and so on.

In what follows, then, mathematical assertions will be understood as claims about the future: if certain operations on signs are performed, the result will be as stated.[30] This means that proving assertions, converting them into theorems, amounts to validating predictions about future configurations of signs: what the mathematical Subject as sign-manipulating agency of the Code will experience if the procedures in question are correctly carried out. Observe that this characterization of assertions differs markedly from the view that they are items of possible "knowledge" in the accepted sense of this term—namely, propositions that are true descriptions of some determinate subject matter that exists independently of the knowing subject and in advance of any claims about it. On the contrary, the understanding here is that mathematical assertions concern what the Subject will find to be the case about his or her own signifying activities. Insofar, then, that "knowledge" is an appropriate term for a validated mathematical assertion, it has to be taken as knowledge about and inseparable from the constitution of this Subject. A fuller sense of such a reflexive, non-epistemological way of discussing assertions will emerge after we enlarge on the status of the Subject as an idealization.

How, then, do thought experiments persuade us to accept the validity of an assertion? The assertion makes a prediction about the Subject, in response to which the thought experiment furnishes the Subject with a scenario of what he/she would experience, a scenario of imagined actions acted out by the Subject's proxy, the Agent. This can only impinge on and carry any persuasional relevance for the Subject by virtue of the likeness between Subject and Agent: only on the basis of the recognition "it is like me" is the Subject in a position to be persuaded that what happens to the Agent in its imagined world mimics what would happen to the Subject in the actual or projected world. But it is just this recognition, this judgment or affirmation of sufficient similitude, that the Subject cannot articulate, since to do so would require access to an indexical self-description necessarily denied to any user of the Code. It is in the metaCode, rather, that an external description of the Subject—through its status as idealization of the Person—is available and that the claim of a sufficient similitude between Subject and Agent, imaginer and its imago, can be articulated.

The hierarchy of semiotic agencies here, from imagined Agent to imagining Subject to indexically conscious Person structuring a thought experiment is isomorphic to that on which any dream rests: the Agent maps onto the figure dreamed about, the Subject the dreamer dreaming the dream, and the Person the dreamer awake, consciously interpreting and recognizing the dream. In the case of dreams the necessity for this third layer occurs because the dream-code—the language of experience for the dreamer—is restricted in various ways, not least by a lack of the ability to recognize the dream *as* dream, which makes it impossible *for the dreamer* to articulate the dreamer's kinship to the imago he or she dreams into being. Likewise, the restrictive nature of the Code in which the Subject operates, in particular its lack of indexicality, prevents the mathematical Subject from articulating the status of *its* created fiction. Of course, unlike the sleeping variety, mathematics deals in *waking* dreams, ones which answer to an explicitly formulated assertion, a prediction consciously in place before the advent of the thought-experimental dream experience which constitutes its validation.

One sees now why, in the case of Peirce's description of thought experiments, the need for the third, interpretive layer distinct from that of the Subject and Agent evaporates. In the face of Peirce's implicit assumption that this whole business of self-examination, observation, and reflective abstraction is carried out in the domain of natural language, thus giving the imagining self automatic access to the indexicality found there, the opposition between dream experience and dream experience *as* dream experience cannot be made. In other words, for Peirce the difference between language and metalanguage has been elided, leading at least in the case of mathematical thinking to an illegitimate coincidence of Subject and Person. A third layer present and operative in his characterization can nonetheless be retrieved: the missing Person in Peirce's picture is none other than the authorial Peirce himself; the one behind his text advocating the thought experiment and urging it—persuasively—upon the reader as the universal means of self-persuasion.

The standard picture of a proof offered in mathematics differs markedly from the one being outlined here. It makes no mention of dreams, imagined actions, fictions, and thought experiments, but instead talks of a logico-deductive chain, a sequence of logical moves, each one either given as an initial posit or implied by such a posit or a previous move by virtue of some rule of inference. One starts from the premises and follows, step by step, through the sequence, giving logical assent to each move until one arrives at the conclusion, which is the assertion that was to be validated. Such a picture of proof as logical deduction is both accurate and, in a radical and essential way, incomplete. Certainly for a proof to occur it is necessary that a logically correct list of entailments with appropriate ending, manifest as a sequence of moves made by the Subject in accordance with the logic sanctioned by the Code, be exhibited. But it is by no means sufficient.

This is because any interpretation of what it means to prove or show or demonstrate or validate an assertion requires more than this formal description can deliver; requires, in fact, that *conviction* take place. It is perfectly possible, indeed very common as all mathematicians know, to agree with and fail to fault the logic of every step in a sequence offered as a mathematical proof without experiencing any sense of conviction. And without such an experience, without a felt sense and a correlative understanding that the reasoning achieves what it is supposed to achieve, a sequence of steps fails to *be* a proof; it fails to be the means of persuading anybody of anything and instead remains as a formally correct but inert string of logical moves. Thus, if one identifies a proof with the logico-deductive object that constitutes its formal appearance, then one's description remains entirely inside the Code; whereas, as we have indicated, the whole nature of validatory reasoning—its grounds in terms of the likeness between Subject and Agent and its object as a prediction about the Subject's future encounter with signs—spills outside the Code into the metaCode.

Put another way, proofs are not proofs until they have been accepted as such by the mathematical community—an acceptance impossible to activate except through the metaCode. It is no doubt a recognition of this extra-formal non-Coded feature of proofs that has prompted some mathematical logicians to recognize proofs as "intensional" objects to be treated within metamathematical investigations as radically different from the purely extensional entities—sets—that are supposed to make up the objects of mathematical discourse.[31]

The interpenetration of Code and metaCode, the way a proof moves inside and outside the Code, is evident as soon as one admits the difference between assenting to the individual steps of a proof and cognizing the whole sequence of them. Presented with a new piece of reasoning, a proof they have not seen before, mathematicians will—often before anything else—ask for and try to find the *idea behind the proof.* They will want to know the principle, concept, or underlying story that organizes the separate logical moves presented to them into a coherent, graspable whole, into a unified argument which, once grasped can be compared and acted upon—simplified, replicated, modified, generalized, and so on—to create other proofs. As Peirce observed in relation to any kind of logical argumentation, every argument rests on what he called a *leading principle*, which is to be distinguished from both the premise and the conclusion (if one were to avoid an infinite regress) and which carries within it the burden of the reasoning in question. The underlying story—a semiotic construct not to be identified with the individual steps, but to be understood rather as immanent within them—is mathematically paramount: once one understands *that*, the whole persuasional structure of a proof can begin to emerge.

Let me summarize the model as presented so far. The scheme is a tripartite one: an Agent, automaton with no capacity to interpret or autonomously signify, performs imaginary acts on ideal signifiers; a Subject, embodied intentional being, manipulates signs whose interpretations bear on the imagined worlds within which the Agent moves; a Person, with access to the appropriate metasigns, observes and articulates the Subject's signifying activities in relation to the Agent. The contention has been that in any process of mathematical reasoning all three agencies are simultaneously present and coproductive. In the metaCode the underlying story organizing the proof steps is related by the Person (the dream is told); in the Code the formal correctness of these steps as a deductive chain is worked through by the Subject (the dream is dreamed); in the subCode of the imago the particular operations that witness these steps are executed by the Agent (the dream is enacted).

The most visible and palpable of these agencies is the Subject— the writing/reading embodied sign-user who is addressed by the Code. From this corporeal Subject come, projected in one direction, a skeleton diagram, the fictive imagined-into-being Agent standing for the Subject, and in the other direction, the Person, a shadow presence constituting within natural language the Subject's otherwise inexpressible spatio-temporal location and tracing—via a prior intention and narrative frame—the Subject's posterior manipulations of signs. It is in this last sense that the phenomenon of metaCode shadowing, mentioned earlier as a striking feature of the zero sign, comes into

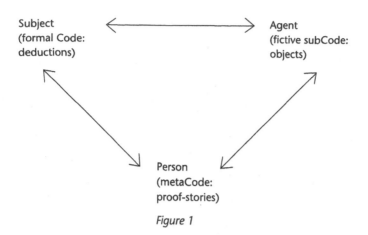

Figure 1

play—less spectacularly than for "0" to be sure—for all mathematical signs when they become part of a process of validated reasoning. The synchronous copresence of Code, metaCode, and subCode, their simultaneous participation within any mathematical act, is repeated diachronically in their genesis. It is impossible, in fact, to write any (but an artificially truncated, impoverished, and mystifying) history of mathematics except as a circuit of changes, initiations, agendas, influences, and reactions that originate variously from the purely formal problems and symbolic developments of the Code, from the cultural institution and legitimation of the Agent, and from the desiderata and demands of instrumentality made through metaCode and maintained there by technoscience and commerce.[32]

We might diagram this summary in the form of a circuit, as shown in Figure 1. The indirect path here, from the Code of pure logic through the metaCode to the subCode of mathematical objects, constitutes the persuasional loop through which any thought experiment works to validate an assertion. Because of this loop, mathematical objects cannot be disengaged from the discursive activity that demonstrates their properties and gives them meaning, with the consequence that the very constitution and ontological status of these objects is inseparable from those cultural and historical practices—the motivation, legitimation, transmission, institutionalization, and implementation of mathematics—that traverse the metaCode.

The triadic scheme here suggests the basis for a unified critique of the three standard accounts of mathematics—namely, Brouwer's intuitionism (mathematics as Kantian constructions prior to any signs—"languageless activity"); Hilbert's formalism (mathematics as game with uninterpreted symbols—"meaningless marks on paper"); and Frege's Platonism (mathematics as discovery of truths, of thoughts "timelessly true, true independently of whether anyone takes [them] to be true" [1967: 29]).[33]

On the one hand, each of these standard accounts denies any constitutive role or mathematically significant presence to the meta-Code. In the case of intuitionism the Agent, too, would appear to be denied, leaving only the (transcendental Kantian) Subject within whose individual, but universal, intuition mathematics has to be constructed. For formalism the relation between Code and subCode reduces to a pure manipulation of material signifiers, while Platonism recognizes this relation exclusively in terms of an investigation by "mathematicians" of a priorly existent, timeless domain of extradiscursive objects. On the other hand, each account can be seen to achieve its understanding of mathematics through an occlusion or a reductive simplification of one of the principal constituents—signifier, signified, sign-using subject—of the semiotic model presented here. Thus, intuitionism totalizes the importance of the signified as mental construction, accepts the necessity of a "creative subject," and repudiates the role of the signifier. Symmetrically, formalism has no trouble acknowledging a (game-playing) Subject and articulates its whole analysis on a repudiation of the signified in favor of a totalized signifier. And for Platonism, human signifiers relate to timeless signifieds, and there is no separately identifiable Subject, only an infinitely idealized Agent and a subjective, psychologized—and so mathematically illegitimate—Person.

Now, against such criticisms, it is undeniable that mathematicians *do* think and perform activities entirely in their heads, *do* scribble and manipulate inscriptions according to formal rules, and *do* discover and discern relationships that, in some undeniable sense, are waiting there to be found. But, according to the model presented here, one has to say the following. The things inside their heads—thought experiments—are impossible, literally unthinkable, without being written and, conversely, are unwritable without being cognized: thinking and writing being cocreative and mutually constitutive. Mathematicians manipulate signs they themselves have imagined into potential existence. This means that what is involved in mathematical activity is not Hilbert's "meaningless marks" but *signs* that have interpretations intended by mathematical Subjects. And the so-called truths discovered by mathematicians are not pre- or extra-human timeless thoughts, but predictions about the Subject's future engagement with signs: predictions each of us—insofar as we occupy the discursive position of being Persons which is offered us by the "I" of natural language—is persuaded by mathematical arguments to accept.

Finally, let us return to the question of the relationship between the status of real, empirical experiments of physical science and *thought* experiments of mathematics. From one direction—that of its concrete interpretation and intended use—the move mathematics makes from actual, instantiable numbers to the manipulation within abstract arithmetic of their fictive counterparts exactly parallels the shift from real to virtual, from the doing/seeing to the imagining of doing/seeing, accomplished for physical science by the "technology of virtual witnessing." Indeed, Shapin and Schaffer's description of that technology cited earlier, involving "the production in the reader's mind of such an image of an experimental scene as obviates the necessity for either direct witness or replication," whereby persuasion operates to convince its practitioners of the validity of substituting imagined actions for real ones, is essentially that of a thought experiment as understood here. The two technologies coincide. What, then, is the difference between thought experiments and real ones? How are they, in practice, related?

Before I can respond to this, it is necessary to recognize a certain difficulty which must then be put aside for the moment: namely, to what extent can mathematics itself be considered as an experimental science? Or, if this seems too shocking—is it not an item of faith that no empirical consideration can impinge on the truth or falsity of a mathematical proposition?—to what extent within mathematics is there a discernible split between the theoretical and the experimental cognate to that in physical science? This question, raised in a crude form by J. S. Mill in his account of arithmetic but long dormant, has been reactivated with the advent of programmable computers—not so much in their role as calculating machines (though this too is vital and part of the issue) as in their ability to aid the construction of proofs. Is a proof which relies essentially on a computer program— which is too long or complex to be understood by a mathematician otherwise—as legitimate a construct as a conventional proof, or is it, relying as it does for its validity on an implementation within a physical machine, in some way empirically or experimentally tainted? There is no simple or definite answer. Rather, the question points to the need for an examination of what the mathematical community does or doesn't want to understand by "mathematician," or, in the present terms, what is or is not to be included in the cognitive repertoire of the "Subject"—a point we shall return to in Chapter 4.

Observe that the diagram in Figure 1 summarizing a mathematical thought experiment could just as well be a model of a typical *Gedankenexperiment*, such as Einstein's image of lightning striking two ends of a train simultaneously. In place of the mathematical Subject would be the corresponding semiotic agency in the relevant science: the one who uses and is addressed by that science's discourse. In place of the mathematical Agent would be this scientific Subject's proxy which, in the example here, would be a figure traveling on the (imagined) train. And, correspondingly, there would be the scientific metaCode which would be the source and repository of all the theorizing, explanatory means and those implicit assumptions of natural language commonly made or sanctioned by that scientific community. The principal difference between the resulting picture and that of a *real* scientific experiment would not be in the triadic structure—any real experiment would be validated by just the same form of persuasional loop through theory—but in the real and actual as opposed to imaginary and symbolic nature of the objects. But even here there is no absolute separation: throughout physical science these real objects will be permeated by and predicated upon mathematical idealities. Indeed, so deeply embedded is mathematics within the very structure of technoscience, as tool, epistemological principle, mode of description—as language—that a complete separation between mathematical or scientific thought experiments on the one hand and real scientific ones on the other is not feasible. One technology of persuasion is folded inside and inextricable from the other.

Evidently, then, any account of the difference between scientific and mathematical experiments has to put into question how what is "real" is to be opposed to what is "imaginary." Now, what is imagined for the purposes of scientific thought experiments has to be a state of affairs that, though it need not "occur in nature at all," as Kuhn expresses it (1977: 240)—need not, in other words, ever occur in fact—*could*, in principle, occur. Thus we retrieve, by way of a large but necessary detour, exactly that difference, that gap between seeing/doing and the imagining of seeing/doing encountered earlier. We return, in other words, to the difference between "direct witness or replication" of an experimental scene and the "production in the reader's [here the Subject's] mind of such an image" of this scene, between "in practice" and "in principle" activity, the explication of which was recognized earlier as a fundamental problem—a problem that arises in a particularly sharp and urgent form in the case of mathematical thought experiments and that now needs to be addressed.

The model presented here displays mathematics as wholly thought-experimental, its entire discourse consisting of a web of waking dreams, conscious fantasies that operate on a separation between activity performed or performable by the Subject and that which is dreamed, imagined to be performed by the Agent. Thus, to push matters further, what has to be elaborated is the difference between these semiotic agencies. The Agent as an abstraction ("skeleton diagram") of the Subject is both a truncation—dealing in uninterpreted signifiers against the Subject's traffic with signs—and an idealized simulacrum, a sufficiently similar image of the Subject to serve as its proxy. The question, then, from one side is what is *sufficiently* similar, how ideal must "ideal" be? And from the other side: is it not possible to go beyond sufficiency, to so extend the process of idealizing a putative proxy that the result is *excessive*, is dislocated from anything that can be given corporeal sense? Would not such an idealization's behavior, its effects and the consequences of its imagined activities, fail to impinge in any materially useful way—in particular, in that relation to technoscience which gives mathematics its value and importance—on the Subject whose proxy it is intended to be?

$\ldots 4$

Lacking a good French name for its devices, IBM turned to Professor J. Perret of the Sorbonne, who suggested the name "ordinateur." That was a theological word, which had fallen into desuetude for six centuries. "God was the great *Ordinateur* of the world; that is to say the one who made it orderly according to a plan."

· Ithiel de Sola Pool, *Technologies Without Boundaries*

To identify the role of human agency in the making of an item of knowledge is to iden- tify the possibility of its being otherwise. To shift the agency onto natural reality is to stipu- late the grounds for universal and irrevocable assent.

· Steve Shapin and Simon Schaffer, *Leviathan and the Air-Pump*

W e have spoken of the passage of Subject to Agent, of Person to Subject, as abstractions. It would have been more precise to have spoken of forms of principled "forgetting" to describe the processes of reduction and truncation whereby the Person (metalingual, indexical/reflexive, involved in arguments via metasigns) gives rise to the Subject (lingual, collaborative, involved in deductions with signs), who in turn produces the Agent (sublingual, mechanical, involved in actions on signifiers). Baldly, the move from Person to Subject is organized around the forgetting of indexicality, and the move from Subject to Agent around the forgetting of sense and meaning.

Now complementary to this sort of principled amnesia is another sort of abstraction: what is not forgotten can be retained in fictive, idealized form. Though frequently co-occurrent, forgetting and idealizing are two theoretically distinct sorts of abstraction, as can be seen in numerous particular cases. Forgetting occurs when all information of a particular kind or mode about a situation is suppressed. Thus, mathematicians might replace a geometric object, a spatial, visually presented labyrinth, for example, by a purely arithmetical object— such as an incidence matrix of numbers representing nothing about the labyrinth except its abstract pattern of intersections. Whereas idealization occurs when aspects of a situation are replaced by purified constructs, here characteristics are not omitted but attenuated to some fictional limit. For example, mathematicians studying the pendulum replace the rhythmic displacement of a swinging object by a fictionalized point on a two-dimensional curve satisfying a differential equation.

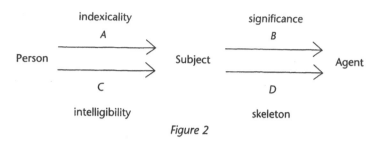

Figure 2

For ease of reference, as well as a certain theoretical suggestion to be taken up later, I shall call these abstractive processes *functors*—the first kind "forgetful" functors, the second kind "limit" functors—and represent them by arrows, as shown in Figure 2. Thus, *A* and *B* denote the forgetful functors representing, respectively, the loss of indexicality incurred in the production of the Subject and the loss of sense in the passage from sign to signifier undergone by the Agent. By contrast, *C* and *D* denote limit functors representing processes of idealization. Our immediate aim is to explicate these latter functors: first *D*, showing how the Agent is constructed as an idealized Subject; and then *C*, showing how the Subject is likewise an idealization of the Person.[34]

The Skeleton Functor: Agent as Ideal

The Subject, a cognitively idealized model of the Person in a way yet to be explored, is a corporeal sign-user, an embodied reading/writing manipulator of materially presented signs. How far can this corporeality be attenuated to produce an acceptably idealized Agent? Faced with the Subject's mortality, one could grant the Agent not one but several lifetimes in which to execute actions such as counting. But why not a thousand or a million lifetimes? Restricted to the energy resources of a single body, one could make available to the Agent the entire mass/energy of the planet. But why not that of the solar system or the galaxy or the visible universe? Bound by the memory/storage capacity of devices in our world, one could allow the Agent to utilize galaxy-wide structures for the storage and retrieval of data. Then why not the observable cosmos? Against the familiar sources of noise, random error, and arbitrary distortion that surround us, one can allow the Agent to operate in deep space isolated from such disturbance at near zero temperatures. But isolated to what extent and how near to zero? Evidently, the difficulty is not what would constitute the attenuated physicality of an idealized Subject, but where one stops the process: *how* ideal? Stopping at some cutoff point, a designated limit of the kind just indicated, has little purpose to it unless the limit springs from or is regulated by some intrinsically plausible, nonarbitrary characterization of what an Agent is supposed to be—and just such a nonexternal characterization is being sought here.

Of course, it is possible not to stop at all. This is in essence the strategy of infinitistically conceived mathematics. And, indeed, for such mathematics the question here is artificial and unreal. The whole issue of idealization and limits is preempted by the prior assumption of a natural infinity of whole numbers, an endless progression given and waiting somehow for the Subject at the outset; thus, to avoid the absurdity of a proxy not being able to reach or count as far as the Subject "knows" the numbers to extend, the classical mathematical Agent must itself be "infinitary." Such an Agent, if it is to perform the operations imagined for it by the Subject, will not merely exhibit a weakened or diminished physicality, some idealized version of the Subject's corporeality, but will possess no physical presence whatsoever. It will be a ghost: disembodied and unrelated to energy needs or spatio-temporal location or the effects of noise.

But, since the whole point of the present investigation is to put the naturalness and transparent endlessness of the whole numbers into question, conceiving of the Agent in this transcendental way not only begs the question; it is totally unrevealing. The very initial acceptance of a potential or actual infinity of objects outside human agency precludes any answer to the question of how mathematical thought experiments might achieve conviction about such numbers. For, as we have seen, conviction works through a likeness between Subject and Agent: to be persuasively effective the Agent has to be like enough to the Subject for the latter to be able to accept it as a proxy yet sufficiently different from the Subject to make thought experiments worth thinking.

The crux is how similar, how different? Not to ask this question, then, is to rest content with an Agent who exceeds any meaning one can attach to the idea of a simulacrum and thus to leave the status of the infinite—what it means and how we are to think it—untouched in heaven, inaccessible to human understanding. Why, in other words, should an embodied Subject, whose identity, conscious existence, and signifying intentionality are inseparable from (facilitated and consti- tuted by) its corporeal being in the universe, create a totally disem- bodied figure to act as his or her proxy? How are the doings, the imagined actions of such an Agent, supposed to engage with the Subject's time-bound, energy-consuming, material presence? Should one not, on the contrary, respect the evident sensoriphysical being of the Subject, the always-given corporeality, and require that the Agent exhibit it in suitably idealized form? Can we not, to put it positively, insist that, in order for any such simulacrum to take part in *persuasive* thought experiments, it must be bound—however remotely—to that same material universe which we, and therefore all potential mathe- matical Subjects, inhabit?

After all, we dream up an Agent and imagine its activities via a thought experiment in order to replace the *doing* of these activities. And we do this because performing such activities on a real-time or real-space basis would be impractical, physically difficult, onerous, irk- some, or otherwise undesirable. In mathematics there is pressure to totalize this replacement: to have the dreamscape extend as far as possible, the attenuation of the Subject's physicality be maximal. But against this, if an Agent is to be created which respects the embodied status of the Subject, then such a figure cannot be allowed to per- form any action that is not capable—potentially—of being realized, of being materially instantiated and made actual within the physical universe inhabited by the Subject. Going beyond such theoretical, this-universe actualizability, asking an Agent to perform *inherently* nonrealizable actions, is to invoke a being who moves in a universe other than the one we—and all conceivable mathematical Subjects— occupy, a being able to "act" and to "count" (assuming meaning can be given to these verbs) outside the regularities of space, time, energy, materiality, and entropy theorized by physical science as governing "acting" or "counting."

Thus, for example, a counting process relying on faster-than-light signaling or requiring more energy to execute it than is usably available in the entire universe has to be disallowed, however "natural" it might be to formulate or operate with the idea. Likewise a machine required to operate at a rate beyond that permitted by the second law of thermodynamics or within tolerances entailing a degree of descriptive accuracy exceeding that prescribed by quantum uncertainty, or to produce, store, and retrieve more individuated signifiers than there are isolatable bits of matter/energy in the universe—all easy enough to specify—would be disallowed. Moreover, in practice, there is no reason to suppose such limits would be independent of each other: as a counting process was indefinitely extended, the limits might well interact and potentiate each other. One can imagine having to supply so much energy, for example, to a process of counting taking place in some region of the universe that consequent changes in the region's space-time, as well as the effects of the entropy produced, would affect the rate, reliability, integrity, and ultimately the identity of the counting process itself. In short, any degree of idealization, however Subject-remote the resulting idealized Agent is, is permitted—provided only that it does not transgress the limit of realizability. The envelope, then, of the Subject's attenuated corporeality is simply every imagined action for which it is not impossible that it be instantiated and become actual.

This criterion, though excessive and unrestrained to the point of absurdity in relation to any realistic or practically conceivable limit, nonetheless immediately replaces the disembodied Agent, the ghost figure of infinitistically conceived mathematics, with a figure whose imagined actions are extrapolations of those in the embodied world of the Subject, an imago whose activities are constrained, however circuitously or remotely, by the limits of the actual. The result, as we shall show in detail in Chapter 5, is a new picture of counting and number governed by a non-Euclidean arithmetic, based on a radically different interpretation of what it means to iterate, to "go on and on." We can see the outline of this if we consider the standard counting process that the Subject imagines being carried out. The Agent is required to start with a signifier, the result of making some—any—mark, "1", and then instructed to repeat the action to produce 11 and then go on doing this, so as to produce the sequence 1, 11, 111, 1111, 11111, . . . , in the familiar way. But now, instead of interpreting this sequence of iterated acts as going on unchangingly, homogeneously, and "endlessly," we must see it as progressing in the face of intrinsic bounds, limitative forces coming into play that ultimately annihilate it.

The insistence that the Agent's actions be realizable is in effect a replay of the issue behind James Clerk Maxwell's well-known attempt to circumvent the second law of thermodynamics. Maxwell suggested a thought experiment centered on the fact that different molecules in a gas of uniform temperature move at very different velocities:

> Put such a gas into a vessel with two compartments [A and B] and make a small hole in the AB wall about the right size to let one molecule through. Provide a lid or stopper for this hole and appoint a doorkeeper very intelligent and exceedingly quick, with microscopic eyes, but still an essentially finite being. Whenever he sees a molecule of great velocity coming against the door from A into B he is to let it through, but if the molecule happens to be going slow, he is to keep the door shut. He is also to let slow molecules pass from B to A but not fast ones. (Quoted in Rothman 1989: 93)

The doorkeeper—Maxwell's Demon—presents a direct challenge to the second law which, among other consequences, forbids the possibility of a free flow of heat from a cool to a hot body. For, as Maxwell immediately pointed out, by its actions "the temperature of B may be raised and that of A lowered without any expenditure of work, but only by the intelligent action of a mere guiding agent." In the century or so since he first proposed it physicists have given various counterarguments to Maxwell's thought experiment. The following, given its final form by Leon Brillouin, is generally held to settle the matter. Brillouin noted that any observation of the physical world rests on the acquisition/handling of information. Information comes in the form of bits which have to be *measured*. By working out the minimum energy needed to measure a single bit of information, Brillouin was able to show that the Demon will always use at least as much energy to carry out its observations as its door-keeping activities manage to create.

Maxwell's Demon is evidently a thermodynamic version of the mathematical Agent. By having been made "exceedingly quick, with microscopic eyes," the doorkeeper is offered as a proxy for an idealized scientist. Unlike the Agent, the Demon is described by Maxwell as "intelligent." It is hard to know what this means here, but, at least in Maxwell's employment of him, this characteristic is superfluous, since the observational and executive tasks assigned it are purely mechanical, making the two simulacra entirely parallel. Now the nature of the idealization is paramount here. Brillouin's argument works because the Demon is, as Maxwell insists, an "essentially finite being," by which seems to be intended that it cannot act outside physical law, that what it knows must be the result of engaging—by measurement— with some physically presented phenomenon. Were this not so, none of the counterarguments—Brillouin's or any other—could get a purchase on the Demon's activities. Likewise, and this is the burden of the case presented here, the mathematical Agent's physicality, however attenuated, is decisive. In its absence the Agent is the ghost of infinitistic counting, and there is no more to say. In its presence the Agent, honoring rather than ignoring the materiality of signifiers, is indeed an "essentially finite being," though to avoid begging the whole question of infinity we have not described it thus. The physicist Dennis Gabor encapsulated the failure of Maxwell's Demon by observing, "We cannot get anything for nothing, not even an observation" (1964: 132). Adapting this, one might say here: we cannot get anything for nothing, not even a further act of counting.[35]

Such a reinterpretation of iteration entails that the ideogram of indefinite continuation, ". . .", can no longer be accepted without comment, but has to be redefined: the cumulative effects of realizability oblige us to recognize the demands of this constraint as an internal and constitutive principle of counting at the very outset. Thus, in place of the classical expression 0, 1, 2, 3, . . . , an alternative notation has to be used. We shall signal this revaluation and the limit associated with it by the ideogram ". . . \$" and write 0, 1, 2, 3, . . . \$ to denote the result of counting, of going "on and on," by a realizable Agent. Of course, the question immediately arises as to whether this same revaluation should not also be announced for the individual iterates, the ordinals "1," "2," "3" that figure in the above expression. Must it not be also the case that the notion of "1" suggested here differs from its classical version, that the realizable and infinitistic interpretations of "+1" will have to diverge from the very beginning? If this is so, should we not be obliged to signal the fact and write not "1," "2," "3" but, say, "1_s," "2_s," "3_s," and so on? The short answer is that we should, but this would be cumbersome and, for reasons to emerge, not necessary at this point in the story. The issue is clearly important and requires a more elaborate response; we shall return to it in Chapter 5 after we have set out the structure of the realizable iterates and their arithmetic in greater detail.

The revaluation here means that, from the metaCode—from a position outside the Agent's counting and being imagined/articulated by the Subject—we see iteration itself become increasingly more difficult to execute, losing integrity as the iterates fade out and become less individuated, more inaccessible, indiscernible, finally dissipating into nonexistence at the limits of what can be physically realized in this universe.

To describe counting—the originary and most distilled form of pure repetition imaginable—as a "dissipating" process that runs down to nothing is undeniably strange and counterintuitive, not to say paradoxical, if one sees counting through the optic of classical infinitism. How could it not be? Infinitistically, the whole numbers are always-already-there, ready to mark successive positions on an actually or potentially infinite line or infinitely subdividable interval. If this is so, it makes little sense to talk of the business of "producing" them, of their becoming "progressively more difficult to execute," and even less sense to see the process of counting, which is endlessly continuable by the very conception of these priorly given numbers, "dissipate" into nonexistence. But the presence of paradox evaporates and both these refusals of sense are overturned as soon as one drops out of the heaven of preexisting, changeless forms and recognizes the necessity of counting numbers into being by accepting the need for some agency to construct them into a—suitably imagined, idealized—mode of existence. Observe that this recognition is posed as, and can claim to be no more than, an alternative conception of what is to be understood by the progression of whole numbers, not a *refutation*, were such a thing feasible, of the classical infinitistic interpretation.[36]

The counting agency, the Agent who brings the iterates into being, does so as an automaton, a robot diagram of the mathematical Subject. To understand the Agent in this way as being at the horizon of what it means to be physically instantiated, is to present it as a construct which substitutes for real robots—present-day computing agents—an idealized machine. The Agent is seen as the envelope of all realizable computations circumscribing any process held to be mechanical within mathematics including, *a fortiori*, the process of counting. One can, therefore, compare this universal limit construct, the *Limit machine* let us call it, with the universal machine formulated by Alan Turing to serve as the abstract mathematical diagram of a computing agent. Turing's project, to model "mechanical process" as this concept occurred in relation to mathematical thought, was conceived entirely within classical, infinitistic terms. He identified an abstract "computation" with the input data and internal configuration of some specifiable Turing machine—an idealized reduction of a typewriter—that would execute it. The business of *carrying out* the computation, of articulating what or who might, in principle, be involved or called upon to "physically" run or be able to run his ideal machine—the ideal typist as it were—formed no part of Turing's picture. Or rather, no problematic part, since the classical context in which he operated already internalized an infinity of numbers immediately accessible to a—necessarily disembodied, dephysicalized—counting agent; hence, for example, the lack of any qualms involved in Turing's assumption of an *infinite* memory tape as part of his machine's specification.

The ideal and reduced typist—the computing agency, unmentioned by Turing, who would run a Turing machine—is of course the transcendentally given, infinitistic agent of classical mathematics.

According to the present account, in which only a proxy for the Subject with an idealized but not nonexistent body is permitted, such an agent appears as an inaccessible and unactualizable ghost, a being whose very failure to be corporeal precludes it from ever impinging on, let alone being identified with, the Agent who runs the Limit machine.

Investigating the physically realizable—the identity, nature, scope, character, workings, and limits of what is and could be actual in this universe—is the business of physics. And, short of an extreme and totalized subjectivism, the answers physics gives to such questions will have empirical, extratheoretical content. It follows that, by inserting realizability into the very foundation of what is meant by counting and number, we are making mathematics depend ultimately on empirical, real-world contingency rather than be a matter, as it is in infinitistic conceptions, of some prior-to-the-facts, purely theoretical givenness. This does not mean, however, that there is any conceptually simple route from physics to the empirical determination of these limits, nor that it makes much sense to ask physics to provide specific information about the fate of iteration as it is pursued by an embodied Agent—to ask, in other words, for physics to give a numerical bound or estimate to the limit, $, of realizable iteration.

There is an obvious reason why this should be so. Certainly, physics wouldn't exist in any form we would currently recognize were it not for mathematics. Thus, there is a kind of methodological trap waiting for any too-speedy expectation that physics be used to investigate mathematics. Moreover, the mathematics on which physics rests—though physics and science generally have no essential need for any infinitary formalism—is formulated in unexamined and uncritically embraced infinitistic terms. It would, therefore, not be coherent to expect a physics so conceived to be able to investigate the basis for an understanding of number—and come up with *numerical* answers— which outright denies these terms. Only a deinfinitized physics could hope to investigate the nature of the bound at $—to estimate its "size," for example—and manage to avoid compromising itself at the outset in this way.[37]

Nor is it a prime consideration for us to ask physics to carry out such a task. It is one thing to appeal to the constraints of physical realizability to establish the *existence* of a limit, quite another to pursue the question of what empirically and numerically this limit might turn out to be. And it is only the former that is required here. Once some limit, however specified, is put into place, a picture of mechanical repetition as a necessarily dissipating process is inescapable, and its radical incommensurability in relation to the classical, infinitistic scheme is ensured.

The question of incommensurability here is in effect a reprise of the difficulties identified earlier in the case of the sorites paradoxes. Recall how such paradoxes arise from an unbridgeable gulf between what syllogistic reasoning, operating within the classical logic of repetition, predicts and what actually happens. The problem seems to be that any item-by-item map from syllogisms to the empirical states of affairs they are supposed to represent breaks down under cumulative repetition. An exactly parallel difficulty blocks any attempt to construct a function establishing an item-by-item mapping of 0, 1, 2, 3, . . . onto 0, 1, 2, 3, . . . \$. Only here, where there are no vague sorites-type predicates involved but only a relationship between two forms of pure iteration, the difficulty is starker and more readily identified. The issue turns not on the existence of a map—one can always find classical number notations to name the realizable iterates—but on the ability of any such map to preserve structure, to be in mathematical terms a *morphism*. What has to be preserved, in any "modeling" or appropriation of realizability which classical mathematics might attempt in order to deny the incommensurability of the two schemes, is internal, arithmetical structure: the map must take classical addition onto its realizable version, classical multiplication onto realizable multiplication, and likewise for exponentiation, hyperexponentiation, and so on. And this, as will become clear, is exactly what is unavailable.

Now arithmetic, as a formalism for the manipulation of the iterates, rests, obviously enough, on the introduction and definition of the arithmetical operations themselves. Introducing these is a discursive move, a bringing into play of certain sign practices, that lies outside the purely mechanical competence of the Agent. Sign practices are the work of the Subject, so that to understand how these practices are supposed to impinge on the iterates brought into being by the Subject's idealized proxy, we have to examine the sense in which the Subject itself is an ideal being.

The Intelligibility Functor: Subject as Ideal

What is the cognitive repertoire of the Subject? What sign-using, computational, linguistic capacities and intentional givens are we obliged to attribute to this mathematical idealization of the Person? In the present account the Subject has at least to be able to read/write a syntactically heterogeneous range of inscriptions, recognize assertions, respond to inclusive imperatives, initiate the process of counting, invoke an Agent to act as its proxy, imagine the activities of this Agent in the narrative frame of a thought experiment, observe these activities, and follow the logical pattern of a proof which details them—all in such a way that, applied to assertions, produces conviction. Any truly adequate picture of this repertoire would have to explicate how these capacities—to signify, recognize, think, repeat, observe, imagine—were the mathematically idealized attributes of corresponding features of the Person: a formidable and unmapped program in the history and philosophy of mathematics that lies well outside the present account.[38] Clearly, we must limit our expectations and restrict attention to certain essentials.

For example, considered purely on the level of sign use, in its role as an idealized reader/writer of materially presented mathematical inscriptions, certain features of the Subject seem immediate and straightforward. Standard mathematical practice stipulates (or, rather, implicitly assumes) the Subject's ideality to be such that he or she be free of error, misreadings, boredom, fatigue, memory loss, and misperception, insofar as these affect the reading/writing and manipulation of Code determined signs. Even on this level, however, there are problems: thus, behind this readily assumed ideality is the banal but essential and inescapable question of *size*. How much information can the Subject actively survey, remember, store, retrieve, and manipulate? What range of signifiers, how intricate a textual presentation, how many repetitions of the same signifier, how many lines of proof, how elaborate or complex a diagram, how extended a definition can one expect the Subject to make sense of and incorporate into a thought experiment? Such questions—and indeed any others about the Subject's constitution—have received remarkably little attention from the mathematical community and until recently would, in fact, have seemed odd, irrelevant, and at best of no more than tangential importance to the business of doing or even describing mathematics.

It is the advent of computer-aided reasoning in mathematics that has rendered this attitude untenable. What is forced into the open is the question of the incorporation of such reasoning into the mathematical process. Is mechanical augmentation of the Subject's cognitive operations to be permitted as unproblematic, or does it introduce an illegitimately experimental and empirical content into mathematical results? The use of a machine to process a large number of logical cases too extensive to be handled or surveyed by any individual mathematician—most famously in a proof of the four-color theorem which guarantees that only four colors are needed to color a planar map such that no two adjacent regions have the same color—makes it imperative to discuss what is or is not to be accepted as part of the Subject's cognitive repertoire. Though brought into prominence and given a certain urgency by the use of computer programs, the sort of problem and concern indicated here is not new: both Hume and Descartes worried about the question of following very long mathematical proofs. And though their solutions to the difficulties are not of direct relevance here, the forgotten nature of their concerns about the excessive length of ratiocinative chains only underscores the lack of contemporary attention given to the issue of determining what a cognitive baseline for the mathematical Subject might or should be.[39]

We shall not pursue the topic of computer-aided reasoning here, but instead focus on the issue that would undoubtedly underpin any discussion of it, namely the process of idealization intrinsic to the Subject's ability to repeat. Plainly, the capacity to be able to do again with signs what one has already done, to iterate a symbolic activity, is a characteristic and central item in the Subject's reading/writing and imagining processes. Nowhere is this more so than in arithmetic. For the Agent, whose activities form the material of the Subject's arithmetical imaginings, repetition consists of the purely mechanical process of iterating the successor operation, S, that is, of adjoining some fixed signifier to obtain the sequence 0, $S(0)$, $S(S(0))$, $S(S(S(0)))$, and so on, where, of course, the continuation sign "and so on" is to be interpreted in terms of realizability. So that writing $S(0)$ as 1, $S(1)$ as 2, and so on, we retrieve the previous expression of this sequence as 0, 1, 2, 3, . . . \$.

In manipulating and reflecting on these numbers, the Subject is—from the moment the usual arithmetical operations are introduced—involved in a seemingly endless hierarchy of repetitions, and repetitions of repetitions. Thus, from the successor operation S, which produces each noninitial number from its predecessor, one—that is, the mathematical Subject—forms, in the usual way, the operation of addition by repeating S a total of n times to give $m + n$ as $S . . . S(m)$. Similarly, one obtains multiplication by repeated addition to give $m \times n$ as $m + m + . . . m$ (n times). Exponentiation is then repeated multiplication to give m^n or, in more convenient notation, $m(exp)n$, as $m \times m \times . . . m$ (n times). Likewise one forms hyperexponentiation as repeated exponentiation to give $m(hyperexp)n$ as m with a sequence of n exponents. And so on.

But what, in terms of the Subject's cognitive repertoire, is to be understood by "and so on" here? Certainly, this repertoire must be rich or extensive enough to contain the means for symbolically processing the results of these arithmetical operations, and it must do this in such a way that the Subject can relate them intelligibly to the sequence of numbers 0, 1, 2, 3, . . . \$. But, to repeat the question, *how many* such new operations can the Subject be expected to handle? How far up the hierarchy of repeated repetitions can the Subject be assumed to cognize? Thus, we have a sequence of operations S, $+$, $\times$, *exp*, *hyperexp*, . . . being introduced, and the question is: how are we to interpret the repetition ideogram ". . ." that regulates the Subject here? Let us write this ideogram as ". . . @," so that the sequence of arithmetical operations is written in the form S, $+$, $\times$, *exp*, *hyperexp*, . . . @. Then our question is: what is the relation between the limits designated by @ and \$?

In the classical, infinitistic understanding there is no way of posing this question, and certainly no discernible problem here. The sequence of arithmetical operations as well as the sequence 0, 1, 2, 3, . . . are standard infinite objects: the latter representing the progression of natural numbers, the former the standard infinite hierarchy of primitive recursive functions defined on these numbers; and their limits, denoted in the usual way by ". . ." are identical. Now, although the present perspective replaces the classical ". . ." by realizable ". . . \$," it might still be possible that a relationship parallel to that in the classical case holds, namely: @ and \$ are identical.

But a moment's reflection shows this cannot be so. There is a fundamental separation between 1, 2, 3, . . . \$ and S, +, ×, . . . @, not marked by their self-evident difference in content but arising from the principles regulating their coming into being as *sequences*, that would be obliterated in such a parallelism: a difference between the mechanical iteration of signifiers by the Agent and the cognitive repetition of sign operations by the Subject. One cannot conflate the products of these two radically separate semiotic agencies: unlike the Agent—pure automaton—what the mathematical Subject does must be intelligible, intersubjectively interpretable in terms of signs. So that, granting this difference, the question of @ as a limit remains: what is there to guarantee the preservation of intelligibility? Why should one be able to suppose that *duplicating an intelligible act will itself turn out to be intelligible*? To ask about "@," then, is to ask about the limit of intelligible repetition.

In the case of nonmathematical situations the experience of such a limit of intelligibility, what one might call a repetition fade-out, is very familiar. In speech one has embedded clauses within embedded clauses and so on which rapidly becomes incomprehensible. In literature and film one has the device of flashbacks, then, more difficult to control, flashbacks within flashbacks and, then, on the edge of sense, flashbacks within flashbacks within flashbacks. In theatrical representations a play within a play within a play, as in Corneille's *The Illusion*, seems as far as one can go without losing the ability to intelligibly interrelate dramatic levels. Somewhat differently, but related to the perceptual psychology of these phenomena, there are several analogues of Abraham Maslow's rule, according to which "the closer a need comes to being satisfied, the larger an increment of additional gratification will be required to produce the same satisfaction," as well as various, somewhat unsuccessful, attempts in economics to formulate laws of diminishing returns and diminishing marginal utility.[40]

On extramathematical grounds, then, there is a substantial and palpable difference between the passage to the mechanical limit $, the principle of this-universe realizability regulating 1, 2, 3, . . . $, and the much shorter-range phenomenon of cognitive/psychological diminution that is operating in the examples here, where the ability to make active sense of the distinctions between different levels of repetition disappears rapidly. The question becomes: what is the relation between this latter type of fade-out and the fate of the mathematical Subject's repetitive activities? In what sense does the process of repetition which starts from the successor operation and produces addition, multiplication, exponentiation, and so on come up against a boundary or limit, quite distinct (in genesis and scope) from the purely mechanical dissipation governing the Agent's activities, beyond which it is not feasible—intelligible, cognizable—to go?

Each of the arithmetical operations after the successor operation S comes into being as a repetition of the previous one. And since the repetition of repetition of . . . is ultimately a repetition of what was originally repeated—here the successor function—it would seem that nothing essentially new is introduced: all that occurs is merely the introduction of new names or notations that serve to abbreviate expressions that go back ultimately to the individual iterates in 1, 2, 3, . . . $. Thus, $m + n$ is nothing more than an abbreviation for $m + 1 + 1 + . . .$ (n times), $m \times n$ merely an abbreviation for $m + m + m + . . .$ (n times), and so on—so that what can be said using these abbreviations could equally well be said (at much greater length to be sure) without them.

Now even in the classical case, where there is no possibility of new items coming into being (it is axiomatic that the sequence 1, 2, 3, . . . is closed under $+$, $\times$, exp, . . .), denying any creative or mathematically significant role to the introduction of abbreviation-based operations is not plausible. One knows, since the metamathematical results of M. Presburger and Kurt Gödel, that classically conceived arithmetic with both + and × is radically different from a system with only +.

In the present case, this closure property, as will emerge, fails to hold. As a result, relegating the introduction of arithmetical operations to mere "abridgements of discourse," denying that repetition is the vehicle for conceptual novelty, and claiming that signifiers introduced to abbreviate repetition are not the source of new mathematical substance is yet more implausible. Plainly, if we want to see what is cognitively at stake for the mathematical Subject to generate the progression of arithmetical operations, we must ask why the Subject's discourse needs to be abridged. What is so misleading about the assurance that one can say the same thing "at greater length" in the absence of an operation as one can say more shortly with it? To this end it will be helpful to defamiliarize the arithmetical situation by comparing it with other hierarchies of cognitive levels. Specifically, let us juxtapose the mathematical Subject's arithmetical repetitions with those of other subjects associated with the codes of speech, alphabetic writing of speech, and formal systems, as shown in this tabulation.

hyperexponents	stories of stories of stories	metagenres	metametaproofs
exponents	stories of stories	genres	metaproofs
products	stories	narrative texts	proofs
sums	utterances	sentences	assertions
iterates	meaningful units	words	terms
0	phonemes	letters	atomic signifiers
ARITHMETIC	SPEECH	WRITING	FORMAL SYSTEMS

We are interested in what is involved in a vertical ascent within the columns of the table here. The process at work as one moves upwards is concatenation: items at each level are amalgamated into sequences or chains to form an item at the next highest level. Of course, there are crucial differences of status and character between items at the same level, but for the purposes of the illustration these can be ignored. What is of interest are the vertical commonalities that seem an inherent feature of any semiotic system. Thus, however formally presented, proofs are always specifiable as certain chains of assertions, narratives always presented as ordered linkages of sentences, products as sequences of sums, stories as linear progressions of utterances, and so on. And in all cases our cognizing of an item—the sense, meaning, or significance we are able to attach to it—has to rest on (though by no means be reducible to) this construction from below. The question we are pursuing takes the form: how much further could our table be upwardly extended?

If we respond phenomenologically as cognitive actors, the answer is not very far, if at all. Even as it stands, one—that is, the writer and many readers—experiences difficulty trying to cognize items on the sixth level. Thus, hyperexponents have no agreed upon notation, are difficult to concretize, and indeed play no practical role in our arithmetization of the world. Likewise, stories of stories of stories, metagenres of narrative texts, and proofs whose objects are themselves proofs about proofs have an intangible, unstable quality, and one is unsurprised by their absence from the respective codes. And if the items on level six are on the edge of intelligibility, what of those on level seven? And even supposing one could make usable—repeatable, disseminable, instantiable—intersubjective sense of, say, metametametaproofs by publicly imagining the appropriate thought experiments about thought experiments about thought experiments, what of level eight or nine: how are we to think those? Are they, could they be, intelligible, imaginable thought possibilities?

Evidently, such direct questions about the outer experiential edge of cognizability can only be rhetorical: one can no more exhibit or make manifest such a limit than think the unthinkable or utter the ineffable. What is undeniable, however, is that the attempt to cognize items at ever-higher levels encounters a certain systematic and increasing resistance. The imaginable becomes ever more tenuous and faintly connected to and indistinguishable from what has already been imagined, until it simply disappears. Any limit, then, can only be inferred, identified from below, as the endpoint of a process whose stages, though cognizable themselves, are patently less so as the process is extended.

We shall complete our account of idealization by drawing out certain observations from the foregoing picture.

First, the phenomenon of repetition fade-out exhibited here should not be confused with what is called in French *mise en abîme*: the effect produced when a representation (for example, a painting of a gallery wall) contains an internal copy of itself (it itself appears as one of the paintings on the wall) which in turn contains an internal copy, and so on. What does "and so on" mean here? The spectacle—whatever labyrinths it might give rise to—is that of a mirror-in-the-mirror repetition, a purely mechanical duplication of representations, in other words, an imaged form of *counting*. Classically, the "and so on" of the *abîme* is the *ad*—ghost produced—*infinitum*, the ". . ." of infinitistic mathematics. According to the present model the "and so on" of representation is precisely the passage to the limit of realizability. As such it regulates the fate of the mathematical Agent. But it is the Subject's capacity to repeat its actions on signs and remain intelligible, not the Agent's mechanical ability to act on de-signified ideal notations, it is the embodied, intentional writer/reader, not his or her imago, that we are concerned with here. In other words, the *abîme*'s limit is precisely the mechanical dissipation to nothing signified by $; but the limit whose character is being sought here is rather the passage to @, the cognitive fade-out into unintelligibility.

Second, it should not be thought that ". . . @," interpreted thus as the boundary of the Subject's capacity to intelligibly repeat, is to be understood as a transcendental limit, a universal human cognitive constraint. This would only be a return to Kant's system which would oblige one to see it as fixed outside society and outside language, whereas, on the contrary, it is a product of history and discourse—more precisely, a product of the systems of writing/imagining numbers brought historically into being within mathematics. As we have seen, for the Code of mathematics the cognizable is neither more nor less than the symbolizable, since the inseparability of ideas from their inscription, of signifieds from signifiers, inherent in mathematical activity and manifest as we have seen in the production of a thought experiment, forces one to couple what is imaginable with the inter-subjective production and exchange of written signs. All that is claimed of ". . . @," then, is that, at any given time, it exists as an historically manufactured boundary.

Thus, by way of illustration, consider the moves involved in ab-breviating the individual iterates of the sequence 1, 11, 111, 1111, 11111, . . . \$. As they stand, these iterates iconically image the succes-sive acts of counting (n strokes representing n acts of counting). The introduction of + and × allows the familiar decimal positional notation to replace the sequence 1, 11, 111, . . . 1111111111, . . . \$ by 1, 2, 3, . . . 10, . . . \$. But this too exhibits iconicity at a higher level (n prod-ucts of 10 requiring n positions) leading to the need to introduce exponentiation whereby, for example, 1,000,000,000,000,000 is written 10^{15}. But again iconicity reappears at the next higher level leading to the introduction of hyperexponents; and so on. In each case the Subject introduces a new arithmetical operation together with a syntax based on the principle of a new name made by count-ing constituent signifiers of names at the previous level. Plainly, iconic-ity is always only deferred, never abolished; there will always be, at any historical moment, a level at which iconicity is untouched—and therefore a level at which the relevant signifiers are unreadable—because the arithmetical operation needed to abbreviate and render the manipulation of them intelligible is uncognizable by the Subject.

Third, though the practical relevance of the limit denoted by ". . . @" will emerge in Chapter 5 when we come to consider the par-ticular character of realizable arithmetic, its theoretical import turns always on the manner in which the Subject is an idealization of the Person. For, were we to deny or ignore the existence of @, we would be involved in a paradoxical situation whereby the Subject could attach intelligibility to, could claim to cognize, constructions which the Person, whose stand-in the Subject is, could neither understand nor be in a position to imagine into existence; this possibility contradicts the dependence of the Code on the metaCode and, in particular, the manner in which thought experiments by a Subject allow the Person to be persuaded. Thus, in the context of such experiments one imagines doing things in order to avoid *actually* doing them. But what if these "things" are themselves imaginative acts? Is it possible to imagine imagining *X* without actually imagining *X*? Or to imagine imagining imagining *X* without imagining imagining *X*, and so on? What evidence could be produced to substantiate such putative acts of meta-imagining? And if, as seems to be the case, it isn't possible to even suggest let alone document such evidence, then one is obliged to acknowledge the existence of a limit to what can be Imagined about imagining: which is indeed all that is being claimed for the arithmetical Subject here.

Fourth, the relationship between Agent, Subject, and Person diagrammed earlier in Figure 2 as a pair of linked functorial connections is highly schematic: any entity *Y*—a situation, property, process, object, agency—held to be a model of some comparable entity *X* could be represented formally by a pair of linked truncating and fictionalizing processes:

$$X \underset{\longrightarrow}{\overset{\longrightarrow}{}} Y$$

What makes the schema specific to mathematics is, of course, the content of what is forgotten (indexical presence by the Subject, signifieds by the Agent) and the content of what is fictionalized (cognitive capacity of the Person, embodiment of the Subject). And, though this is essentially all that we shall make use of in the critique of infinity being developed here, a certain qualification needs to be registered in relation to this last point.[41]

The overall picture offered here has been essentially synchronic, mathematics being treated as if it were given in its entirety as an already established discourse without regard for the facts of historical change and the various ways mathematical practice has emerged and been developed. This means that I have not examined, but have taken for granted, the fact that mathematics is a social-historical artifact brought into being over several millennia by a heterogeneous but identifiable community of practitioners. What are here called the Code and metaCode have resulted, as historians of science would be quick to point out, from a complex of cultural pressures, goals, and ideological imperatives that range from the external demands of science, technology, and commerce for a mathematics to serve their (particular and historically various) ends to the institutional, disciplinary, and professional requirements of the mathematical community to serve *its* quite different psychological and philosophical ends. To have entered into any kind of adequate engagement with these issues would have entailed examining mathematical practice in regard to the *socio-cultural* idealization of its practitioners and the way they engage in the transmission, exchange, and legitimation of its texts as cultural objects. Instead I have presented the outline of a semiotics that has confined itself to a very local and persistently focused explication of the principal features of the trio of Person, Subject, and Agent that are irreducibly present in any piece of mathematical reasoning.

Evidently, however, the character of these semiotic agencies and the manner of their connection do not come from nowhere: behind the purely theoretical, semiotic move from Person to Subject to Agent there is historical process. In particular, and less obviously perhaps, there is a back-formation at work from Agent to Subject to Person that reverses their analytical order. Thus, for example, the idea of "two" available to and so part of the *Person's* cognitive repertoire is inseparable from that of "2" elaborated by the Subject in the Code: the Subject-created mathematization of reason, through the application of its formalisms to the world, is a constitutive element of the Person. Likewise the very nature of a mathematical Subject, its possible identity and horizon, is affected by how and where the idea of "Agent" as a mathematical automaton and abstract imago originates.[42] We shall give a fuller discussion of the various two-way traffics involved here—between natural-language numbers and their mathematical counterparts, between Code, metaCode, and subCode, between Person, Agent, and Subject—in Chapter 5.

Finally, observe that the limits charted here, which have involved differentiating between the classically unending *ad infinitum* ". . . ," the iteration limit ". . . $" of this-universe realizability, and the fade-out ". . . @" of cognitive repetition, have arisen not as a result of some prior philosophical decision about the nature of mathematical objects or some external regulative scheme of how mathematics should be practiced, but from considerations that seem inescapable as soon as the text-based, ideogrammatic writing/thinking character of mathematics and of its persuasional techniques is recognized.

This is not to say, however, that the identification of these limits was achieved from some theoretically neutral starting point (were such a thing to be possible) or that once identified these limits have no normative consequences. Far from it. The whole account here springs from a semiotically based refusal to accept the currently available explanations—in fact, lack of explanations—as to how the natural numbers come into the world to be humanly observed and manipulated. And the consequences of this refusal can hardly avoid being normative and revisionary in relation to the infinitistic interpretation that underpins all current understandings of what it means to call the whole numbers "natural." Indeed, as we shall now see, once they are brought together, the two limitative principles of realizability and cognizability allow a model of number and its resulting arithmetic to emerge that is radically unlike and ultimately incongruous with the familiar classical picture of an endless and unchanging repetition-of-the-same given us by Euclidean arithmetic.

. . . 5

Should it not be expected, then, that in the domain of the very large surprises await us. . . . And the possibility is not excluded that the description of the situation may demand essentially different constructions of the very mathematical basis, that is, our assumptions on very large numbers.

· P. K. Rashevskii, "On the Dogma of the Natural Numbers"

I t will be evident by now, without as yet going into the details, that an interpretation of counting as a dissipative process is of necessity in direct conflict with the standard, infinitistic conception of number and its arithmetic—both of which I shall designate as Euclidean. The conflict is located in the divergence between the classical ad infinitum interpretation of the ideogram ". . ." as endless continuation and its realizable version ". . . $" as continuation governed by the limit of realizability. Strictly speaking, we should talk of an absolute difference rather than a "divergence," since the principle behind the sequence 0, 1, 2, 3, . . . is, from the very beginning, radically separate from that producing the sequence 0, 1, 2, 3, . . . $. And we should, then, more properly write this sequence of realizable iterates in a way that signals this, such as $0_\$, 1_\$, 2_\$, 3_\$, . . . \$$ where $0_\$$ signifies the origin of counting, as initiated by the mathematical Subject within the tripartite model set out here, and where $1_\$$ denotes the first iterate, $2_\$$ the second, and so on, produced by the Agent imagined to count by such a Subject. We could then, as we shall want to do, make a comparison between classical i and realizable $i_\$$ and so, on the basis of this, make sense of talk of "divergence" between them. For the most part, however, we can do this without using the $-subscripts, which we shall drop and leave it to the context or the presence of ". . . $" to convey the difference between the two sorts of iterates.

What, on the basis of the geometrical analogy, might one mean by a non-Euclidean arithmetic? Recall that Euclid's axioms for geometry rest on an essential equivocation, a failure to distinguish between axioms as truths about space, and axioms as purely logical or mathematical posits. Before the advent of non-Euclidean geometry in the

nineteenth century these two readings of "axiom" were fused: it was
held as obvious—evidentially undeniable—that the axioms as set
down by Euclid were, whatever their origin in Greek mathematics,
necessarily true and final descriptions of space. The logical consistency
and internal, explanatory coherence of non-Euclidean geometries
allowed axioms as truths and axioms as posits to be separated and the
descriptive status of Euclidean geometry to be properly questioned for
the first time.

The possibility of non-Euclidean arithmetic—whether it can be
formulated and what it might mean—turns (at least within the perspective adopted here) on whether one can treat geometry's relation
to its object as a paradigm for arithmetic's relation to *its* object.
Geometry's object, what it describes, theorizes, or models, is extension in space; arithmetic's object is passage through time. Is there
a legitimate basis for a symmetry of treatment and conceptualization here? On the face of it, at least from an historical direction,
there would seem not to be. Thus, for example, the impact of non-
Euclidean geometries on Kantian arguments for the *a priori* Euclidean
nature of space is well known. One of its forms was a recognition that
Kantian "intuition" of space—far from being apodictic or transcendentally founded—was no more secure than our strongly felt inability to
coherently deny the "truths" enumerated by Euclid's axioms. There
seems not to have been any corresponding philosophical recognition
about our intuition of time: geometry's theorization of space and
arithmetic's of time were treated (outside the interchange of time and
space in post-relativity physics, which is a different matter) as entirely
separate. This was reflected, for example, in Brouwer's attempt, long
after the implication of non-Euclidean geometries had been assimi-
lated, to found his—endless—construction of the whole numbers on
our supposed Kantian intuition of time. But it is this very endlessness
that is in question here, so that making appeals to an "intuition"
founded on it is incoherent; and it is in question in a way that could
hardly be closer, as we shall see, to the geometrical situation that confronted Euclid.

For, as is well known, what exercised Euclid about his own axioms

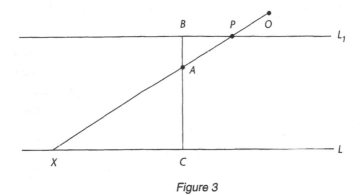

Figure 3

was the status of the axiom of parallels, which asserted that through any point P not on a line L only one line L_1 could be drawn parallel to L. Although the axiom of parallels seemed obviously "true" to Euclid, something about it made him uneasy, and he tried, as did many after him, to deduce the postulate from his other axioms. A troublesome feature of the postulate was undoubtedly its reference to and involvement with the idea of a straight line being prolonged infinitely, or unendingly, far. Not only is this idea implicit in the definition of parallel lines as ones which *never* meet, but it seems implicated, in a less obvious way, in the truth of the postulate itself. For in order to be convinced that no line other than L_1 can be drawn parallel to the given line L through a point P, one draws lines like OPA, as shown in Figure 3, and observes how each such line must meet L at some point X, and so cannot be parallel to L. However, as one moves A nearer to B, the point X recedes unimaginably far along L—the nearer B is to A the more extreme the effect and the more inaccessible X becomes—with the result that the line CB has in some sense to be infinitely divisible. And infinite divisibility was not an idea that Euclid could have been happy to incorporate into his geometry, since this was precisely one of the assumptions that Zeno—whose paradoxes had left Greek mathematics with a lasting fear of both motion and infinity—had reduced to absurdity. In any event, whatever the source of Euclid's unease about the axiom of parallels, a mathematical basis for his doubts was articulated in the nineteenth century when two different and incompatible denials of the axiom were formulated and shown to be internally coherent.

The particular character of these denials and the geometries they give rise to need not concern us here. Our contention is that the cognate in arithmetic to Euclid's axiom of parallels is the ad infinitum principle itself: endless prolongation of counting, of iteration in time, is no less problematic, no more necessarily "true," than unending extension in space. In addition, insofar as the interpretation of ". . ." formulated here as ". . . \$" is intelligible and internally coherent, it will give rise to what can legitimately be called non-Euclidean arithmetics. In other words, once one recognizes the inescapable temporality of counting, then what holds for geometry in relation to spatial prolongation holds for arithmetic in relation to temporal repetition. This entails that arithmetic has to be considered no less a model for the structure of actual, empirical time than geometry is a model for the structure of actual, externally presented space. And conversely, it entails that failing to make this recognition—the position of classical infinitism—leaves our understanding of arithmetical axioms on a par with the classical, pre–nineteenth-century understanding of geometrical axioms as "necessary truths" that manage, in some mysteriously felicitous way, to be at the same time both logically undeniable and true descriptions of actuality.

To make our contention plausible, let us observe first that the currently accepted axioms of arithmetic—for example, Peano's axioms, no less than Euclid's axioms for geometry—are derived from everyday experience of handling and observing physical objects. This experience with ordering and collating collections of objects has given rise to an image of the positive whole numbers articulated in terms of simple and familiar laws, which go as follows. Any number can be reached by counting in units from zero; counting is a uniform, homogeneous, and constant process whose character is independent of where one starts or how far one counts; of any two distinct numbers one will be greater than the other; no matter what order one counts any collection of things the final answer will always be the same; one can always form a number by adding any number i to any number j or equivalently adding *j* to *i*; likewise multiplying two numbers together is always possible and independent of the order in which the operation is performed; any number added to itself or multiplied by itself any number of times will result in a definite and unique number; any set of positive numbers, however specified, has a minimum member; and so on.

The image of number enshrined in these laws has delivered a coherent, robust, and totally reliable account of what is involved in numbering collectivities that has survived several millennia of use from the practical exigencies of commerce, finance, and trade to the technoscientific theorization of physical reality. Manifestly, any proposed redescription of number and arithmetic that did not respect the empirically sedimented "truth" of this picture would forfeit any chance of being taken seriously as an explication of what we call "number." But our belief in the truth and undeniability of the picture grows out of our local—human-scale, middle-range—experience of collectivities. What grounds are there for assuming the picture to be *universally* valid? Why, leaving aside reasons of theoretical simplicity or aesthetically shaped preferences, should we suppose that the laws and regularities of concatenation, ordering, collating, and counting that appear to hold for the numbers of our experience extend to the numbering of all possible or imaginable collections of things as yet unexperienced and perhaps even unexperienceable? What sort of warrant is there for believing that what appears to us as locally so is universally the case?[43]

For geometry this separation between the small scale and the universal, that is, the necessity of distinguishing the local from the global, is entirely familiar: we no longer assume the earth to be universally flat because of its flatness in regions of our immediate experience. Neither do we assume that the space of the universe is described by a geometry which seems to fit the structure of our local experience so successfully. On the contrary, twentieth-century physics has, since the general theory of relativity, incorporated the hypothesis that the structure of space, though it may be reasonably and successfully described as locally Euclidean, cannot be supposed Euclidean on a global scale.

And why should the phenomenon be any different for time than it is for space? What more reason do we have to think that what is locally acceptable (successful, "true," applicable, useful) in the case of arithmetic and temporal repetition will be universally acceptable than we have for geometry and spatial extension? Of course, if we are to respect, as we must, the sedimented, human-scale truths that lie behind the image of number given above, then any arithmetic we suggest must, in a sense yet to be explicated, be *locally* Euclidean. To understand what this means we need to be more explicit about the nature of the realizable numbers and the way in which their arithmetic differs from the classical Euclidean picture.

A brief outline of the ground rules and basic prerequisites for realizable arithmetic is given in the Appendix; here we shall present an even briefer and much simplified summary. The picture that emerges interprets "number" as covering two quite separate regions or types of numeration: the *iterates* 0, 1, 2, . . . \$, which result from a process of counting from 0 by a realizable Agent, and the *transiterates*, a region of arithmetical entities that, though not counted into being, are natural and legitimate candidates for inclusion in any minimally satisfactory model of "whole number."

Consider first the iterates. Their overall structure is determined by the limitation on the Agent signified by the presence of \$. One can imagine a primary kind of thought experiment in which an Agent is dispatched to count up to an iterate k, to actively examine k in some way (e.g., perform some task in relation to it), and then to return with the information about k resulting from such an examination. For small enough k the effect of \$ will not be discernible, and such a task will be indistinguishable from its counterpart within classical, infinitistic mathematics. As k is increased, this will become progressively less true, and for large enough k the dissipative effect of \$, which we can think of simply as the result of a resource depletion, will ensure that no such return is possible. There will thus be a "point of no return"—call it a_0—which, though not itself an iterate, cuts or bifurcates the iterates into two distinct regions.

If we accept the usual meaning of adding numbers—$i + j$ meaning performing the counting operation corresponding to j after that of i—then, in the simplest model of such a thought experiment, one can interpret a_0 in terms of +, in the sense that for iterates i and j less than a_0 the sum $i + j$ is an iterate, but for a_0 less than i and j the sum $i + j$ cannot be an iterate and is not, at least initially, assigned any meaning. Thus, a_0 acts as an *exit point* for +. If we introduce multiplication as repeated addition, then there will be an exit point b_0 for ×, in the sense that for i and j less than b_0 it follows that $i \times j$ is an iterate, but for i and j greater than b_0, $i \times j$ is undefined. Likewise there will be exit points c_0 for exponentiation, d_0 for hyperexponentiation, and so on, for as long as it makes sense to continue the process.

Thus, as one ascends the hierarchy of arithmetical operations—from successor to addition to multiplication to exponentiation to hyperexponentiation to . . .—a descending sequence $a_0 > b_0 > c_0 > d_0 > . . .$ is generated. The understanding of the iterates that emerges is of a series of bifurcations in what appears initially as an unfractured and undifferentiated whole—a structure of cuts, identified as exit points, in the uninterrupted process of counting, each of which is brought into being by the introduction of a new arithmetical operation. The effect of the limit $ is thus not confined to excessively large iterates but, by a process of retroformation, conditions the arithmetical structure of the entire sequence of iterates all the way back to its beginning. This does not mean, however, as we shall see below, that ordinary arithmetical operations on the beginning iterates—elementary manipulation of "small" numbers, with arithmetic confined, say, to addition and multiplication—will automatically exhibit any discernible effects brought about by the presence of $. But before we enlarge on this point, which is essentially the question of what it means to say the iterates are *locally* Euclidean, let us complete the picture of realizable "number" in relation to the integers as a whole.

The existence of a_0 as the exit point for addition arises from the fact that, unlike in the infinitistic picture of iteration, two counting processes, i and j, carried out consecutively, cannot always be combined into a sum $i + i$ interpretable as a single counting. One can, nevertheless, think of the combination $i + j$ as a legitimate arithmetical entity, a kind of spillover from the iterates which, though not count-able into being from 0, has definite mathematical properties. Such an entity would be a *transiterate* in the sense that its arithmetical defini-tion would force it to be larger than any iterate. Thus, we could intro-duce a region of transiterates, call it $(+), consisting of all sums of iter-ates, representing the closure of the iterates with respect to +. In exactly similar fashion, we could introduce regions of transiterates $(×), $(*exp*), $(*hyperexp*), and so on, consisting of all products, expo-nents, hyperexponents representing the closure of the iterates with respect to the operations of ×, *exp*, *hyperexp*, and so on, respectively. The overall picture, then, of "whole number" or "integers" is the com-bination of a fixed counted-into-being domain of iterates and a cer-tain domain of posited, uncountable-from-0 transiterates.

The question arises of how far the enlargements of the domain of transiterates extend. What is to be the interpretation of "and so on" here? From all that has been said, it will be clear that this is the same question as how far the sequence of operations starting with the suc-cessor function and continuing by repetition with addition, then mul-tiplication, and so on, can be prolonged. As we know, the introduc-tion of these operations is the work of the mathematical Subject, so that the question is in fact asking: how far does the Subject's capacity to intelligibly cognize repetition extend? The answer is the limit we have designated by @. Lacking any external determination of $ and @, we can say nothing more specific than this.

Put another way, the phenomena of exit points and definability, and hence the issue of the "closure" of the arithmetic being sketched here, arise not from some externally imposed numerical limit, but rather the reverse: what is to count as "number" will be determined by two wholly internal constraints. First there is the realizability limit $ which yields the iterates and *a fortiori* the transiterates derivable from them. Then there is the cognitive limit @ which regulates the repeti-tion activities of an embodied Subject called upon to manipulate—read, write, process, notate, handle, decipher—materially presented number notations and expressions. It is possible to bring these limits into a unitary description, to see @ as a reflection of $ inside the world of the Subject or to construe $ as the idealized limit of a @-constrained, this-universe inhabiting Subject. But we need not pursue this since, in relation to the present thesis, the outcome is the same: the Subject's manipulation of such expressions and the application of the results of the manipulation to the original sequence of iterates run into real and unavoidable material-symbolic difficulties—difficulties which increase cumulatively as further operations are introduced until they become, as was demonstrated earlier, cognitively insurmountable.

Let us now return to the general question of how and in what sense realizable arithmetic differs from the standard Euclidean conception. What is the fate of the laws that underpin Euclidean arithmetic from the present perspective? Which relations that appear classically self-evident and intuitively undeniable are preserved and which are made problematic once the effects of @ and $ are taken into account? Should one take as self-evident the commutative laws $i + j = j + i$ and $i \times j = j \times i$ for all iterates i, j? Is it obvious that the identity $1 + y = y + 1$ must hold for all iterates and transiterates? On what grounds should one believe that, if y and z are distinct realizable integers, then either y is less than z or z is less than y? Where does one find the evidence that any set of positive integers must have a least element? And so on.

Consider the commutative law which states that $i + j = j + i$ for all i and j. If i, j are less than the exit point a_0 for addition, then $i + j$ and $j + i$ are iterates and the intuitive justification for the equality of $i + j$ and $j + i$—that two different countings of the same collection give the same result—coincides with the classical case. If however i and j are greater than the exit point a_0, then $i + j$ and $j + i$ are transiterates, and the familiar intuitive procedure for comparing them as entities counted up from 0, together with the induction proof which codifies this procedure, is no longer meaningful. If, more extremely, i, j are themselves transiterates, the situation is even more problematic and divorced from the Euclidean picture. Likewise, the commutative law $i \times j = j \times i$ for multiplication admits an interpretation parallel to the Euclidean one for iterates i, j less than the exit point b_0 for multiplication, but cannot be so understood if i, j are greater than b_0; and the

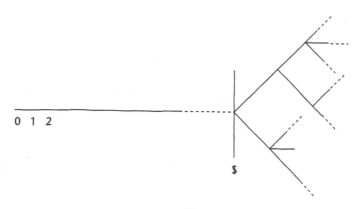

Figure 4

situation is again more problematic if i, j are transiterates. Or consider the law of trichotomy, according to which either two numbers i and j are equal or one is greater than the other, that guarantees total ordering. If i, j are iterates, then the self-evidence of this law—which enshrines the perception that of any two unequal countings from 0 one will be arrived at before the other—is the same for realizable counting as it is for the Euclidean notion. But if by "numbers" one understands not iterates but transiterates, then not only does the standard proof by induction break down, but the picture which sustains it is simply not available, and the only ranking between transiterates that can be taken without qualification is the partial ordering between them that is forced to hold by virtue of their arithmetical construction from the iterates. We might picture the situation as in Figure 4, where the region of transiterates to the right of $ is represented by a schematic partial ordering instead of a line to indicate that total ordering of the transiterates cannot be assumed.

Observe that the point here is not the question of internal coherence or logical consistency attached to various decisions that, for one reason or another, might be made. One can certainly decide, and the move may indeed be perfectly consistent with all that one starts from, that $i + j = j + i$, or that $i \times j = j \times i$, or that the partial ordering of the transiterates be extended to a total ordering. But the ability to make such moves is beside the point: arithmetic here is being construed as a model—more or less adequate, useful, "true"—of the process of realizable counting and cognizable reflection on that counting, not as an aprioristically justified, purely theoretically given *definition* of counting. This being so, the calling into question of classically accepted first-order regularities, such as commutivity and total orderedness, should be seen, not as an attempt to logically deny or in some sense refute these Euclidean symmetries, but as an opportunity to found an alternative, non-Euclidean arithmetical intuition.

The same goes for those second-order aspects of Euclidean arithmetic, namely its satisfying the induction axiom and its closure under the arithmetical operations, whose intuitive "obviousness" is made equally problematic and dubitable by the present account. On the first of these aspects, note that in light of the fact that the transiterates cannot, by definition, be counted into being by a realizable Agent from 0, the integers will not be well-ordered and will not form an Archimedean domain. This means that one cannot assert that for any given y and z there will necessarily be an iterate n such that either n lots of y is greater than z or n lots of z is greater than y. On the second aspect, the failure of classically conceived closure is immediate and, through the phenomenon of exit points, gives rise to the central feature of realizable arithmetic. This is that at each such point one has to decide—that is, the mathematical Subject has to decide—whether to create transiterates corresponding to the sums, products, exponents, and so on of entities beyond the point in question. It is impossible, as we saw, to achieve closure in this way, since the creation of fresh transiterates only exacerbates the problem. The result is that there is no question of "remedying the incompleteness" of the system; one has to accept that the realizable integers exhibit an ever more complex structure of exit points beyond $, and it is only by invoking the limit @ that one can speak of arriving at any sort of "closure."

Enough has been said to indicate the way in which the interpretation of number here differs from its Euclidean counterpart: evidently, realizable arithmetic is radically *non*-Euclidean. But against this difference there is also an identity: any acceptable understanding of number, we agreed earlier, has to be *locally* Euclidean. What is this to mean? In the simplest and most concrete terms, how are the realizable iterates, imagined as counted into being, to be seen as locally identical to, but ultimately different from, the familiar progression of ordinal numbers that we all write down when we count? Granted that $1_\$$, $2_\$$, and so on, dissipate and ultimately vanish at \$, what is to prevent us from nevertheless *identifying* them with the classical ordinals before the effect of such terminal dissipation makes itself felt? After all, how would one notice any "difference" between k and $k_\$$? What arithmetical significance could attach to the idea that 10 and $10_\$$ are not the same? On the contrary, does it not seem that counting to ten on our fingers is counting to ten—whether the counting is prolonged beyond $10_\$$ or 10 by a realizable Agent or by a classical ad infinitum agent able to count forever?

Evidently, a difference in meaning between 10 and $10_\$$—or more generally between the iterates i and $i_\$$ for any i—has to be mathematically manifest, has to correspond to a difference in the behavior of these signs within their respective formalisms. Thus a difference must, in order for it to have mathematical content, exhibit itself as a divergence between classical and realizable laws of arithmetic. It follows that one can only make sense of comparing i and $i_\$$ by first recognizing that the comparison has no absolute status, it cannot even be articulated except in relation to a specified level of arithmetic. The reason for this is that such differences between iterates can only be articulated in terms of the very arithmetical operations being used to detect, compare, and exhibit them.

Thus, at the primary level, consisting of the successor function which generates the iterates but no other arithmetical operations, the difference between the classical and the realizable iterates is exhibited in the presence of \$ itself as an arithmetical limit, since it is only by denying the classical ad infinitum axiom—for any z there exists y such that $y = z + 1$—that realizable arithmetic comes into being at all. The next level, which is where arithmetic proper begins, contains the operation of addition as well as the successor function; here there will be no divergence between k and $k_\$$ provided that $k_\$$ is less than the exit point a_0 for addition; conversely, for $k_\$$ greater than a_0, the arithmetic of sums will not be the same in the realizable and the classical cases. If now the operation of multiplication is introduced, then the point at which the arithmetics diverge will be brought forward to the exit point b_0 for products. At the level that exponentiation is introduced, the divergence point is brought forward to the exit point c_0 for exponents. And so it would go on to higher and higher levels with the introduction of hyperexponentiation, hyperhyperexponentiation, . . . , for as long as it makes sense, according to the limit @ regulating such sense, to do so.

Observe that an essential feature of the realizable iterates, much in evidence here, is the way their global structure emerges as the result of a reflective or zigzag movement—one goes up toward $ to establish properties of iterates down to zero—through an unfolding brought about by raising the level of arithmetical operations. The smaller the iterate k_s the higher the arithmetical level of observation needed to distinguish k_s from k. Lacking any description of the passage to $ which would determine the internal "settings" or "values" at which $a_0, b_0, c_0, d_0, \ldots$ occur, one can say nothing more specific about particular iterates, and our earlier question about the putative difference between the realizable and classical understandings of "ten" can only be answered in a very general way: in the case of a number as small as ten, no arithmetical difference between 10 and 10_s is likely to emerge except by the execution of prodigious computational/symbolic activity on the iterates, of the kind involved in, say, hyper- or hyperhyperexponentiation. At levels below this—restricting arithmetic to repeated addition and multiplication and exponentiation, for example—10 and 10_s will be indistinguishable. Such a restriction is by no means arbitrary or unmotivated in relation to contemporary, classically based numerical practice. The most extravagantly large numbers occur in physics and, with some minor and highly speculative exceptions, theorization of the material universe in physics takes place entirely within a range from 10^{-40} to 10^{80} measured units. Hence, corresponding to the de jure operational closure of realizable arithmetic presented here—the necessity of its boundedness by $ and @—there is, though practically undeclared and certainly untheorized, a de facto closure that operates with respect to the classical employment of "number" in relation to the world—a relation we shall return to in Chapter 6.

We see now that our earlier question of how realizable-arithmetic-as-a-whole is to be compared with its classical counterpart—the question, in other words, of its being *locally* Euclidean—can only be addressed in a general and relativized sense. Thus, in the presence of only +, realizable arithmetic is Euclidean up to a_0 for addition; in the presence of + and × realizable arithmetic is Euclidean up to b_0 for multiplication; and so on. Moreover, the characterization in terms of an ascending series of levels here, though perfectly correct, has to be qualified to avoid a certain spurious accuracy of differentiation: the transitions from one level to another are not abrupt but pass through a finer series of gradations which make it impossible to draw sharp boundaries. (This is obvious even in the classical case where the decision as to how to classify an item such as 10^z for $z = 10^{26}$ will depend on whether one writes it as $10^{100000000000000000000000000}$ or whether one insists on abbreviating the exponent to 10^{26} by using elementary hyperexponentiation). Thus, repeated application of the successor function overlaps and merges with addition, repeated addition merges with multiplication, repeated multiplication with exponentiation, and so on.

Behind this description of the locally Euclidean nature of realizable arithmetic, made relative and general by the question of how *in practice* we are to distinguish a realizable iterate i_s from its classical counterpart i, there is the question of the absolute theoretical difference between i_s and i. The whole conception of realizable iteration entails the denial of any pure, unconditioned identity. The making of an additional mark—the adjunction of 1 to a sequence 1 1 1 . . . 1 of marks—will *always* be conditioned by the size of this already-made sequence. And this is so regardless of the fact that for all sorts of practical and theoretical purposes the effect of this conditioning might be masked and that for small iterates no difference is arithmetically discernable. Indeed, it is this very indiscernability that enables the sorites paradox discussed earlier to get off the ground. But, as will now be evident, the paradox disappears as soon as one replaces the classically conceived iteration it rides on by realizable iteration. An effect of this replacement is to naturalize the familiar qualitative differences between the arithmetical operations and to display them as the emergent effects of repetition. But, and this is the point, such arithmetical repetition is fundamentally the same as that which operates in the world of heaps of grain and heads of hair. It is, in other words, no more to be captured as the result of a self-identical adjunction of a self-identical mark than any other empirically bounded activity. Thus, one can say that, far from being a paradoxical phenomenon of "vagueness" irrelevant to mathematics' exact employment of concepts, the sorites effect is a fundamental and constituting principle of (non-Euclidean) arithmetic.

Observe that the overall structure of the arithmetic presented here can be described in terms of a series of bifurcations involving the breakdown of symmetry. Thus, at $ the symmetries expressed in the identity $1 + i = i + 1$ and in the law of trichotomy—$i = j$ or $i < j$ or $i > j$—can no longer be assumed to hold, and the fundamental division into iterates and transiterates occurs. The operation "+" changes behavior at a_0, rendering the symmetry $i + j = j + i$ problematic; $\times$ changes behavior at b_0, rendering $i \times j = j \times i$ problematic; and so on, for the further arithmetical operations. In each case a unitary classical event gives way to a multiplicity of separate possibilities. Such a state of affairs can be compared with the behavior of certain open dynamical systems. The analogue would be a system whose primary, equilibrium state—where its behavior is "linear"—would be the domain of iterates brought into being by the successor function. The system is observed/investigated by a Subject whose introduction of the arithmetical operations—an input of computational/symbolic activity—alters it and creates nonlinear behavior. The greater the activity, the larger the alteration: each operation enables the system to be driven further from its equilibrium position, so that the region of the iterates where the system's behavior is "linear" will be bounded first by $ and then by the exit points for addition, multiplication, exponentiation, hyperexponentiation, and so on. In each case the boundary between regions will be the place at which the system appears to exhibit a discontinuity in its arithmetical behavior.[44]

There is of course more to numerical mathematics than the whole numbers. The realizable arithmetic outlined here has to be supplemented by the behavior of ratios of these numbers—realizable rational numbers—and whatever structure of realizable real numbers (what one would call the $-continuum) is to be constructed from the set of such ratios. To pursue these matters, particularly the latter, would take us too far afield. Enough has been said, certainly in relation to the implications of denying the ad infinitum axiom, to explicate the claim that the sort of number one would naturally develop on the basis of $ and @ is non-Euclidean. And, going on from this, enough of the resulting locally Euclidean nature of realizable arithmetic has been given for us to be able to sketch how its employment, in place of classical infinitistic mathematics, would have radical consequences for the theorization of the material universe offered by science, particularly physics.

Physics connects numbers to the world through enumeration and measurement. Measurement, whether of time, length, mass, charge, volume, energy, or entropy, is the conceptualized ideal of the practical repetition of standard units: one concatenates "numbers" of seconds, centimeters, grams, ergs, etc., into an ordered plurality to arrive at the measurement of an observable phenomenon. Such a plurality—unless it occurs as an instantaneously presented cardinal (e.g., the number of atoms in a volume of gas or particles in the universe) where no claim is made that any process of consecutive enumeration lies behind the number—has to be the result of a real or imagined process of counting: one is either concatenating units to measure ever-larger or subdividing a unit to measure ever-smaller stretches of time, space, or matter in a straightforward extrapolation from the standard and familiar Euclidean practice.

But as we have seen, such extrapolation is not always possible once certain constraints are respected: lining up units one after the other or successively chopping up a unit into fragments involves an iterative process which is bound by the limits of realizability: any process of concatenation, however presented, can only meaningfully adjoin k units, and any repeated subdivision of a unit can only result in a fragment $1/k$ of the unit where in both cases k is less than \$. There is, of course, nothing to prevent one from talking of k units or $1/k$th of a unit for k greater than \$—that is, for k equal to a transiterate—but one will not then be dealing with a numerical entity that has been counted into being from zero. And the consequence is that the familiar numerical laws which govern counting can no longer be taken to hold for such entities.

To illustrate this, consider the measurement of time as it figures in the program of cosmological physics to describe the ultimate past and future states of the universe. The thought experiment cosmologists ask us to perform is a double extrapolation from the world, the "now" which we, the observers, inhabit. Backward, down to the putative Big Bang Event, time is measured ever closer, in principle arbitrarily close, to the Event at time 0; forward, time is measured up to some distant, in principle arbitrarily large terminus or—in some theorizations—an infinite nonterminus in the future. From the present non-Euclidean standpoint such a thought experiment requires a conception of measurement that is of necessity unreal, literally so in the sense of being unrealizable. First, assuming we are prepared to grant meaning to numerically specified stretches of time larger than any iterate or fragments of time smaller than any reciprocal of an iterate, there is no warrant for supposing that such quantities of time are totally ordered amongst themselves. Second, even this sort of assignation of meaning is unavailable beyond a certain point: a stretch of time "larger" than any transiterate or a fragment of time "smaller" than any transrational (the reciprocal of a transiterate) will be without meaning. And this latter restriction will operate no matter how large the "value" of \$ or that of @ is taken to be and so how far we augment the transiterates, since no such augmentation will give rise to arbitrarily large numbers or produce numbers arbitrarily close to 0.

This means that the demand for arbitrarily small or large *measurable* time, time larger than or "after" the transiterates or smaller than or "before" the transrationals, is without meaning. Such "time"—in the sense that it can be connected to the domain of the iterates by any thinkable sequence of arithmetical operations—is cognizable, if at all, only in nonnumerical language. In this understanding the so-called origin of the universe and likewise its so-called end cease to be "events" extrapolatable from or measurably related in time to any now-event through the apparatus of number. They become instead certain kinds of horizons marking the limits of any possible numerically presented description. And what goes for time also goes for measurability in space; there will be, therefore, no numerical sense that can be attached to an "arbitrarily large" extension or distance nor to an "arbitrarily small" region of space. There is, moreover, nothing special about the role of the putative Big Bang Event here. The same analysis in terms of a horizon holds for any event situated at a supposed *singularity*, whether this is interpreted as a point at which physical parameters become "infinite" or as a point on the edge of the universe where any light ray must terminate, since implicit in either definition is the requirement that the point in question be approached by a process of arbitrarily close approximation—which is precisely what is prohibited according to the present account.

The horizons referred to are horizons of (numerically presented) information. They occur as an inevitable consequence of the nature of realizable number and not as a result of any experimental inadequacy, assumptions within theoretical physics, or the like. Thus, in a physics governed by an arithmetical apparatus which conforms to the limitative principles presented here, any putative event—a supposed singularity for example—which is not realizably accessible to an observer as the actual or potential origin of a measurement system will be surrounded by a zone of numerical ignorance. In the approach to such a zone, measurements will diverge increasingly from their classical counterparts as the numbers determining them cease to obey the familiar arithmetical laws, and within the zone itself—that is, "beyond" the horizon—no numerical description will be available.[45]

Such a description of a "horizon of information" as a function of the limit of what can be iterated and transiterated into being is specifiable and makes cohesive sense from *within* realizable mathematics. What would be the result of trying to describe it externally, from a classical, infinitistic point of view?

Classically, the mathematical description of the continuum along which all physical measurement takes place is a dense-in-itself, complete, ordered field; in relation to this the realizable continuum will appear as essentially incomplete, a totally discontinuous array of numbers full of gaps and holes. What would be exhibited in realizable arithmetic, for example, as the failure of density—between any two realizable rationals there is not always a third such—would appear classically as a simple absence of easily specified real numbers. What would be manifest as a limit—the boundary between iterates and transiterates or between realizable rationals and transrationals or the limit of realizable numbers themselves—would have to be presented classically in terms of certain real numbers that functioned as externally originated minima, that operated in other words as more or less coherently theorized, more or less specifiable numerical *quanta*.[46]

Now whether such a classical interpretation of realizable measurement is sustainable is by no means clear, since the idea of numerical quanta finds no recognizable purchase within an infinitistic understanding of the continuum. And though it is not necessary, nor in the end especially helpful, for the development of realizable mathematics to pursue such an interpretation, it is of interest to record a de facto embrace by physics of such a generalized quantization. Thus, modern physics, which since the advent of quantum mechanics has accepted the quantization of energy in the form of the quantum h of action, of the order of 10^{-27} erg seconds, considers, in addition, certain other specific constants that are, according to Planck, "independent of special bodies or substances, which necessarily retain their significance for all times and for all environments, terrestrial and human or otherwise, and which may, therefore, be described as 'natural' units."[47] Thus, in approximate terms, the Planck time of 10^{-44} seconds is put forward as the minimum conceivable or theoretically meaningful instant of time, similarly the Planck length of 10^{-33} centimeters, and the Planck temperature of 10^{-32} degrees K. One should note that unlike the quantum of action, these supposed minima have nothing in the way of an experimental or theoretical infrastructure to justify conceiving them as true "quanta," and their status as such is accordingly unresolved within contemporary physics. From the present perspective one can speculate that their presence within physics (and hence ultimately an explanation of them) is a manifestation—expressed entirely within the scheme of classical infinitistic mathematics unsuitable to them—of precisely the sort of limit phenomena operative within realizable mathematics.

. . . 6

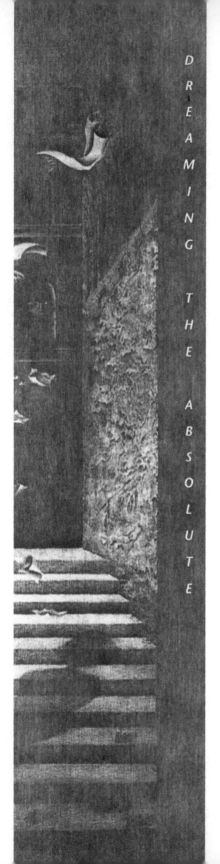

DREAMING THE ABSOLUTE

Let me repeat now that I have reached the end, what I said at the beginning: man would sooner have the void for his purpose than be void of purpose.

· Nietzsche, *The Genealogy of Morals*

The human intellect does understand some propositions perfectly, and thus in these it has as absolute certainty as Nature herself. Those are of the mathematical sciences alone . . . in which the Divine intellect indeed knows infinitely more propositions than we do, since it knows all.

· Galileo, *Dialogue Concerning the Two Chief World Systems*

The purpose of the present essay has been to put the idea of the mathematical infinite into question. In order to do this, to ask what the infinite is and how we are to think it from a position that does not beg the very questions to be investigated, I have insisted that before all else mathematics has to be understood as a language, a discourse with written signs, a manipulation of ideograms inseparable from the making of certain kinds of waking dreams or thought experiments. But there are two other dimensions of mathematics which, though less pertinent to the principal concern here, need to be acknowledged. Mathematics would not command the sort of attention it does and would not occupy the privileged role accorded to it in Western culture were it not a prodigiously useful formalism. And no understanding of it would be adequate that failed to recognize mathematics as a kind of fantasy action, an elaborately ramified series of symbolic games played with numbers, points, spaces, and other such special, peculiarly mathematical objects. I want now to tie the three dimensions of mathematics—the discursive, the instrumental, the ludic— together so as to put the question of infinity into a wider context. Each of them will be seen to rest on a certain exchange, a back-and-forth movement between two poles—signifier/signified, discourse/ world, and writing/imagining—which, though analytically separate and operating over different terrains, are mutually dependent, co-creative, and co-occurrent, neither one having privileged sense, origination, or priority over the other.

Before looking at this exchange at work in the instrumentality of mathematics, observe that from the Platonist/realist perspective of the majority of mathematicians and scientists the usefulness of mathematics appears as mysterious and inexplicable. As a widely quoted and approved formulation by the physicist Eugene Wigner has it, "The enormous usefulness of mathematics in the natural sciences is something bordering on the mysterious, and there is no rational explanation for it. It is not at all natural that 'laws of nature' exist, much less that man is able to discover them. The miracle of the appropriateness of the language of mathematics for the formulation of the laws of physics is a wonderful gift which we neither understand nor deserve" (1960: 2).

How does this miracle come about? Why should the timeless truths discovered by mathematicians pursuing their own exclusively abstract, internally motivated, and nonempirical interests provide such an apt description of the empirical world? What *is* the cause and the source of the "unreasonable effectiveness of mathematics"? What indeed? Short of invoking a deity (the source of the miraculous "gift") responsible for the creation of both mathematics and the world— essentially the response of Galileo, Newton, Leibniz, and many after them—what could provide the basis for the "profound mathematical harmony of Nature," as the physicist Roger Penrose recently put it? How does this correspondence, which seems so wonderfully pre-established, between mathematical discoveries and the structure of physical reality come about? The difficulty addressed by such a question is evidently the result of a re-inscription, on the level of instrumentality, of the fundamental epistemological obscurity of Platonism —namely, how does an embodied, earthbound, knowing subject come to be connected to and make discoveries in the realm of eternal, absolute, and prehuman truth, a realm which, as the Platonists never tire of proclaiming, is/was/will always be "out there"? But awestruck wonder at mathematics' capacity to be so uncannily applicable to the world, though no doubt reinforcing the hunger for the numinous that Platonism itself so evidently caters to, misses the point about how mathematical objects in fact come into being. On the contrary, matching slogan to slogan, one can say that mathematical objects are not so much "discovered out there" as "created in here," where "here" means the cultural circulation, exchange, and interpretation of signs within an historically created and socially constrained discourse.

Instrumentality works through a twofold movement between mathematical signifieds and the world. In one direction the world is manufactured and theoretically constituted from these signifieds: it is manifestly the case that mathematical concepts map out, create, model, articulate, and provide metaphors, metonyms, idealizations, grids, frameworks, and diagrams for the way we organize, perceive, and come to understand many of the various realities we inhabit. According to this, what we call the "world," whether this is the techno-scientific theorized world of material process or the commercial world of circulating money and its instruments, is in a vast and continuing debt to mathematics. Certainly, neither of these two worlds that enclose us is even cognizable outside the mathematical signifieds used first to "describe" them and then to service, maintain, and change them. The world's content as a conceptualized product, then, is indeed given shape and form through mathematical signification. But mathematical signs do not fall from the sky, ready-made and pre-adapted to fit a world that knows nothing of them; they have themselves been extracted from whatever world mathematics has found and changed over the past several thousand years. In the other direction, then, mathematics is in an ongoing debt to the world. Not only are its overall history, its range of particular concerns, and the direction and development of its research programs influenced by numerous social, material, and psychological factors external to itself—that observation is a commonplace of all but the narrowest internalist understanding of the subject—but the actual content and logical shape of mathematical signifieds themselves owe their origin to empirical, material features of the world. Once the originating effect of these features is taken into account and seen as folded in and constitutive of the very abstractions, vocabulary, relations, and objects that we recognize as mathematical, the long-standing effectiveness of mathematics ceases to be surprising. In fact, it's not the successful application of mathematics that is "unreasonable"—it is, on the contrary, its absence that would be irrational.

Indeed, why should one be puzzled by the utility of geometrical or arithmetical discourse, why think they perform miracles upon the world, when their objects are themselves idealizations and truncations of real-world regularities—distances, positions, angles, collations, partitions, matchings, countings, orderings—and when their theorems result from thought experiments designed to preserve the relations between these regularities? Why in fact be anymore puzzled than by the success of the myriad thought experiments each of us performs daily through the employment of natural language to move around, categorize, remember, impinge on, and anticipate the world of material, embodied existence? And what goes for geometry and arithmetic goes, more indirectly to be sure, for topology, algebra, calculus, analysis, and the rest of mathematics: each has arisen out of, and controlled and organized its abstractions in the light of, empirically originated patterns, processes, and regularities.

Of course, to say this is not to rest with any reductively simple account. A huge program of detailed work and many subtle questions arise about how mathematics regulates these abstractions, about what the nature of the idealizations and truncations performed on the world is, about what it means for thought experiments to "preserve" real-world regularities in such a way that these experiments turn out to be useful. Thus, as was evident in our earlier discussion of the sorites effect, any too facile or uncritical acceptance of such "preservation"—in that case assuming the predictive invariance, the "truth"-preservation, of classical, syllogistic logic under arbitrary iteration—leads to absurdity and paradox. But these and similar questions ask one to examine the appropriateness, usefulness, and viability of this or that manner of application. One asks such questions as: Does two-valued Boolean logic adequately model the implicational structure of quantum mechanical statements? Is linear mathematics appropriate for the description of certain dynamical systems? And so on. One does not by such questions, however, put into doubt the empirically derived basis for the application of mathematics to the world.

Consider next the double movement between signifier and signified threading through mathematical discourse. Throughout our explication of mathematics-as-language in terms of the manipulation of signifier/signified couples, we have emphasized the interpenetration and mutual entanglement of these two sides of the sign. Mathematicians (that is to say, mathematical Subjects) scribble down what they think and think about what they scribble, and they do this in such a way as makes it artificial to accord one of these activities any primacy or originating status—of a genetic, conceptual, psychological, or logical kind. The relation is rather one of mutual indebtedness. Mathematicians in their ongoing practice interweave the two, so that what becomes of interest about signifiers is their capacity to engender imagined scenarios and what is essential about signifieds is their intersubjective control and regulation by written inscriptions. And in doing so mathematicians are merely repeating, in the small scale, the larger-scale interweaving that operates at the level of historical change. Thus, new signifieds appear on the historical scene with their meanings determined, at least in the beginning, solely by systems of signifiers. This is obvious in the case of complex numbers which arise from the intention to assign "numerical" solutions to certain equations such as $y^2 = -1$. Less evidently perhaps, it is true again of the system of signifiers—zero together with the familiar positional notation—whose introduction brought about large-scale changes in what was signified by number in Western mathematics that had been unthinkable— because unwritable—before their introduction.

In all such cases the nature and status of what is newly signified depends in turn on yet earlier such interweavings of signifier and signified, and so on. From this it follows that, contrary to Brouwer's and Husserl's intuitionism, there is no prelinguistic mathematical meaning, no pure primary intuition, no origin of geometry or arithmetic waiting to be expressed, no freestanding signified not already caught up in mathematical signifiers. And likewise, contrary to Hilbert's formalism (and indeed those post-structuralist accounts of discourse which seek to give absolute primacy to the signifier), there is no such thing as a pure signifier, no "meaningless mark on paper" in relation to the mathematical Code, and so no coherence in the attempt to understand mathematics as the manipulation of such marks prior to or independent of any possible interpretation.

Now this mutual dependence of writing and thinking takes place through the active presence of the mathematical Subject, the discursive agency required and in turn facilitated by the Code, and it is around this agency that the double movement of mathematical play takes place. For it was, as we saw, precisely in the two-way traffic between the mathematical Agent, who executes actions imagined for it by the Subject, and the Person in the metaCode, for whom such actions are simulacra of the Subject's own embodied movements, that the persuasive structure of mathematical reasoning as waking dream, and so of play, is realized. Neither the Person interpreting such dreams nor the imago-Agent called into existence to enact them can be assigned any primacy. Not only does their relationship through the Code imply that each is subject to the interweaving of signifier and signified just observed, but in addition neither exists except in relation to the double movement between the cultural world (domain of the Person) and the signified (that which is idealized and forgotten by the Agent) that constitutes the historical source of mathematics' instrumentality. One can note in passing here that it is by virtue of these interconnections that the aesthetic dimension of mathematical play— elegance, generality, brevity, subtlety, concision, and economy of means, expression, and content—enters into its application to the world. Contrary to the conventional impulse to separate the hard zone of the functional and utilitarian from the soft domain of beauty, value, and style, in mathematics they are fused at precisely the point where the world—already and ineradicably a mathematized domain— exerts a selective pressure on the mathematics that will best, most aptly, most beautifully, most concisely fit it.

The activity of play, the traffic of imagination between Person and Agent, by no means results in a free, unlimited field of waking dreams available to the Subject. This is because the movement rests on an oscillation—not confined to mathematical signs, but having a particular force for them—between the empirical and the ideal: any signifier appears as both a *token*, an individual, empirically presented ink mark (say), and as a general, idealized, non-materially presented *type*. Only by being inscribed and concretized can an abstract type have any signifying presence; and only in relation to a general pattern of recurrence is it possible to talk at all of a "token." In mathematics, whose objects have an entirely ideal status, the nature of the process of idealization assumes a decisive importance. As we have seen in some detail, what does and does not count as an acceptable idealization is inseparable from the question of what constitutes the "finite" and so of what meaning is to be attached to the "infinite." And this question depends in turn on that of limits: of what is and is not to be accepted as an ideal simulacrum of the embodied mathematical Subject, and on what might or might not be imaginable and cognizable by this Subject in relation to its own repetitions.

There is, then, a more complex interpretation and a more demanding arena of "play" operating here than the mere working out of certain rule-bound games, such as chess (to cite the standard example). The psychiatrist Donald Winnicott has made such an interpretation central to his account of human development. For Winnicott *play*—a deep and universal human activity—occurs in the space between "external or shared reality and the true dream" (1971: 47) and is originally the means by which an individual self negotiates its own boundaries, the divisions between what is inside (the domain of "magical control") and what is outside (the repudiated world, the "not-me"). Being neither objective nor subjective but "transitional" or "intermediate" between the two, and challenging both by a kind of deliberate courting of danger from within their limits, play becomes, for Winnicott, the source of all experienceable novelty and cultural creativity. Making allowance for the fact that in mathematics one has, not an individual self engaged in a psychiatrically motivated mission of survival, but the mathematical Subject—a semiotic agency made available by the Code—engaged in the dreaming of its own numerical boundaries, Winnicott's interpretation is both apt and suggestive. Certainly, in this functional and adaptive sense of play, the waking dreams through which mathematics is staged appear as its neverending attempt to think its own limits, its attempt to go to the logical edge of itself: hence the intense delight in mathematical theorems that are conceptually outrageous, unbelievable, contrary to intuition, excessive, puzzling, "almost false," and so forth; and hence the corresponding anxiety and despair when these limits are transgressed and mathematicians are confronted with contradiction and paradox.

The back-and-forth actions operating within the discursive, instrumental, and ludic dimensions of mathematics work through a certain temporality, a passage of time inscribed within the nature and structure of the movement itself. And this in two ways.

The more obvious and richly layered of these is the cultural temporality of the Person and the metaCode. It is through this that the history of mathematics as language, as a development of ludic forms, and as instrument—encompassing mathematics' relation to the world and its symbolic self-elaboration into Code and metaCode—takes place. The manner in which mathematical objects are abstracted from the world, transformed by the triad of Person, Subject, and Agent through thought experiments, and then returned as technoscientific or commercial applications to a retheorized world to initiate a new cycle means that the history of mathematics-as-instrument and of mathematics-as-sign are in the end inseparable.[48]

Within this large-scale historicity there is the empirical real time of the embodied Subject and, in relation to this, there is the imagined, ideal time of the Agent. For Winnicott "playing is doing" and, as he emphasized, *"doing things takes time."* Certainly, as a reader, writer, creator, decipherer, manipulator of mathematical signs—whose signifiers are presented only and always in their embodied forms as empirical tokens—the mathematical Subject cannot but do things that take time. Indeed, it is precisely because this is the case, because the Subject's embodiment imposes limits (of space, matter, and energy as well as time) on his or her capacity to use, manipulate, and play with actual, empirically present signifiers—to count them into being, for example, in "real time"—that the whole apparatus of thought-experimental reasoning with its conception of the *imagined* manipulation of signs is introduced. What the Subject cannot count into being within its own real time the Agent is allowed to accomplish within the imagined, idealized time put at its disposal. And it is in the nature of this disposal, the nature, that is, of the idealizing relationship which imagined time bears to the Subject's time, equivalently the relationship between empirical and ideal arithmetic (since arithmetic theorizes counting as progression in time), that the question of "infinity" was brought in this essay to a point of explicit choice and decision. And that decision was about what, if anything at all, should constitute an acceptable idealization of the Subject's physicality in the case where the counting is endless.

But any unqualified talk of decisions here is artificial. Once the discussion about infinity is allowed to arrive at this point, there is no real choice to be made: the whole question of what is an "acceptable" idealization has already been preemptively nullified. This is because a choice between idealizations is not present nor conceptually available within infinitistically conceived mathematics. For infinitistic mathematics the numbers are not engendered, they are not obliged to be brought into being by any process, but are before us, already existing in their potential or actual entirety, and any distinction resting on the comparative difficulty of engendering them by one sort of Agent as against another is obliterated. The result is that one has classical infinitism, whose conception of imagined time is transcendental—being in no sense an idealization but rather an outright rupture of the fabric of real time—whose conception of an Agent (were it to analyze matters thus) is a disembodied ghost and whose theorization of time is orthodox, Euclidean arithmetic. Against this one has the Agent of realizable, non-Euclidean arithmetic and a conception of idealized or imagined time as the limit of any embodiable form of temporality— the limit, that is, of the time that the mathematical Subject (and any reader of this text, regardless of whether or not he or she accepts the opportunity afforded by the Code to become such a Subject) inhabits.

In terms of a single idea, then, the response to "infinity" developed here can be summarized as hinging on the issue of *embodiment*. More precisely, it hinges on the insistence that the corporeal mathematical Subject—and his/her ideally embodied Agent—be inserted into the foreground of what we mean by counting and hence "number" and hence what is meant by mathematics itself. One might ask for the reasons behind this. How has it been possible to describe mathematics in such a way that human agency, subjecthood, and corporeality have been so thoroughly occluded? Why has this seemingly banal insistence on the "body that counts" become necessary? And, going on from this, why, at this particular historical juncture in the development of mathematics, two-and-a-half millennia after its classically formulated origins, should the question of the one-who-counts assume, for the first time, such a pressing and vital importance?

Evidently, the Platonistic understanding of mathematical objects as timeless partakers of Truth has no place for human creation or intervention, and its whole ethos could hardly be more explicit in the matter. And even when Platonism as a philosophical system is rejected outright, as it is by present-day constructivists, Kronecker's attribution of the integers as the work of God is still uncritically embraced. Behind both these refusals to allow any role to an embodied human agency in their pictures of number, one can discern a certain ideological profile, the sedimented numerical version of a hierarchy recurrent in Western culture that ranks soul over body, psyche over soma, mind over matter, spirit over flesh, thought over action, ideal over real, mental signified over material signifier, and so on. The very conceptualization of "number" in classical Greek thought was articulated in terms of a particular version of this opposition: the valorization of *arithmetica* (numbers as the proper objects of contemplative thought by philosophers; theoretical objects that appear as ideal, perfect, heavenly, and eternal) over *logistica* (numbers in their practical, empirical, and calculational aspect as the work and servile concern of slaves). This opposition was given material, external form in the period from the third century B.C. to the early Renaissance in Europe through the separation between the writing of numbers in terms of Roman numerals and the practical calculation with them using the abacus.

And, as one might expect, all theoretical and philosophical accounts of number inherited from the classical period have placed arithmetica over logistica no less totally and persistently than theories of the sign have privileged signified over signifier and corresponding conceptions of "person" or "life" have valorized mind, soul, thought, and psyche over their lowly, embodied opposites. The result is that both the mathematical Subject and *a fortiori* the proxy Agent—the one-who-counts—were subordinated to the point of being written out of Western culture's picture of "number" from the very beginning. Nor is the principle of valorization here confined to a pure Platonism: cognate to the ranking of arithmetica over logistica is the Aristotelian elevation of *episteme*, the domain of theoretical and speculative knowledge, over *techne*, the world of the practically and empirically knowable. Since the Renaissance, however, the two oppositions have enjoyed different historical fates: it was precisely against the claims of the Aristotelian Schoolmen that they could theorize Nature in the language of episteme, using the preexisting vocabulary of essences, perfect forms, and heavenly bodies, that Galileo instituted the techne-based apparatus of empirical investigation and measurement which founds modern scientific investigation. That his conception of number nonetheless replicated the primacy of arithmetica over logistica testifies, among other things, to the more deeply embedded rhetorical, ideological, and ultimately theistic structure of the latter opposition.

But this divergence between an episteme overturned by a techne-formulated science and a virtually undisturbed arithmetica-dominated conception of number now looks to be over. The advent of computer science—imminent since the introduction of zero-based positional numerical notation into Western mathematics abolished the abacus-based separation of writing and calculating; explicitly prefigured as a theoretical possibility by Leibniz; attempted in relation to practical schemes by both Pascal and Charles Babbage; and realized technologically in its present form in the wake of the twentieth-century theorization of mechanical process within mathematical logic—has put paid to the subordination of the empirical/algorithmic to the purely theoretical and contemplative understanding of number.

Indeed, as the contemporary manifestation of logistica, computer science signals the instatement of the slave—the one who counts—onto the mathematical scene. An essential part of this instatement is the recognition that any act of counting/calculating, whether by fingers, abacus beads, written marks, or electronic pulses, requires energy, space, and time for its realization. As such, the emergence of computer science represents a large and growing de facto challenge to the subject matter and internal organization of traditional, infinitistically conceived mathematics fostered by arithmetica. At the same time computer science threatens to supplant physical science as mathematics' prime external source of problems and abstractions. Responsible already for the creation of new fields of mathematics, electronically founded logistica has introduced radically new ways of handling and hence cognizing number. Thus, one has fractal geometry, whose development is hardly thinkable except through the use of computer-generated graphic images—which, compared to the manual production of geometrical images, are nothing less than a new mode of mathematical writing. And, closely connected to fractals, in their reliance on recursively iterative procedures, one has chaos theory and the modern investigation of dynamical systems in terms of discrete time states—a manner of study impossible to develop without the computer simulations of such states.

The emergent technology, then, of electronically presented signifiers, from fractal images through phase-space diagrams of iteration-based models to the representation of virtual realities, is an exemplary instance of the double movement described above. One moves from mathematical signifieds to the world to a redrawn mathematics, whereby the computer, a technoscientific product of mathematical thought, impinges on and begins to determine the "internal" growth of mathematics itself. Reading the formation and development of mathematics in terms of such an autogenic feedback opens up new vistas—not least the possibility of understanding the emergence of a genuinely empirical/experimental mathematics.[49] This has many radical consequences, one of which is the impact on mathematics' image of itself as a purely theoretical discipline whose results are in principle incapable of being affected by any "real world" facts and contingencies. In particular, it allows a modified version of Galileo's question to the Aristotelian Schoolmen to be put to contemporary infinitistic mathematicians. Galileo asked: By what principles should we accept, without an examination of the matter, that there are no mountains on the moon because the moon as a heavenly body has to be perfectly spherical? Why should the empirical world of physical phenomena conform to a prior Aristotelian metaphysical theorization of it? Likewise, in light of the above, one can ask: Why should numbers behave according to a priorly conceived, classical schematization of them—a theoretical scheme that assigns them an ideal and perfect status that is an inescapable consequence of Aristotle's ad infinitum principle? On the contrary, is not such a theorization every bit as mysterious, theo-logically self-confirming, and unreal as Aristotle's essences and heav-enly perfections? Contemporary logistica, founded on the proposition that counting—computing, reckoning, manipulating, calculating—is as much a part of the physical world as the mountains on the moon Galileo was at such pains to have witnessed, cannot but introduce the physicality of the one-who-counts, the slave proxy for the Subject, onto the mathematical stage.

And, as we have seen, by an argument based not on computer science but, quite independently, on the exigencies of logical persua-sion and the nature of thought experiments, the existence of moun-tains on the perfect moon of Euclidean mathematics—the systematic failure of classical arithmetical laws and symmetries, the breakdown of the transcendentally ideal intuitions about iteration that underpin them—is a coherently cognizable possibility, a consequence of recog-nizing the embodied Subject and his/her mechanical proxy to be the prime movers in the process of bringing the numbers into being.

However, notwithstanding this independence, there is an undeni-able convergence of interest between, on the one hand, the concerns and theoretical underpinnings of contemporary logistica and, on the other, the semiotic model of mathematics and the prospect of realiz-able, non-Euclidean arithmetics that it makes available.

For the present, such a convergence—which, being speculative and programmatic, can only be glimpsed—prompts two brief com-ments. First, the operative understanding of mathematics that the model rests on—the construal of mathematical texts as fabrics of in-junctions, imperatives, orders, commands, and exhortations addressed either directly to the mathematical Subject or indirectly to the Agent via the mechanics of a thought experiment created by the Subject and interpreted by the Person—suggests an obvious reformulation. One could redescribe the whole enterprise in terms appropriate to an abstractly conceived logistica, whereby any mathematical text would be construed as nothing other than an abstract program, a pattern of instructions to be executed by an Agent operating an appropriately specified theoretical computer. Of course, the operations performed by this Agent would be *imagined* in the now familiar way and, insofar as they were arithmetical calculations, would be carried out on the Limit machine, the realizable analogue of a Turing machine mentioned earlier. To say more about the nature of such an abstract Agent and the theoretical device it operated, one would have to explicate the way in which the device contained internal analogues of the Subject and Person without whom the Agent's activities would be merely formal, and in practice uninterpretable, manipulations.

Second, and here the focus is not what the Agent does but the Subject's capacity to track what it is claimed the Agent does, there is the issue of computer-aided reasoning. Crystallized in arguments that are too long for a naked—computerless—mathematical intelligence to check that the proof does what its proponents claim, the question, as we saw before, is really about what is to be included in the cognitive repertoire of the Subject. And as such, there is no way, other than a normative decision on the part of the mathematical community, to decide it. Evidently, the basis for such a decision is inseparable from the larger question of what meaning and status the mathematical community is to assign to empirical/experimental mathematics. And this in turn raises the very issue that has been central to the present essay—namely, in what sense and by means of what types of justification are we to maintain that the numbers be understood as "ideal," nonempirical entities?

The critique of classical infinitistic mathematics, and the replacement—in effect the de-writing—of infinity presented here, can be articulated as an instance of a simple but far-reaching methodological rule which one might call the principle of *epistemic foreclosure*. According to this, some conceptual item, some previously knowable, sayable, and unproblematically thinkable *X*, hitherto obvious and transparently natural, is denied, to be replaced by a new *X* whose introduction only makes sense in the presence of a new cognitive agency. In our case *X* is the classically understood ideogram ". . ." of infinite repetition, and the denial is predicated on the introduction and foregrounding of the corporeal mathematical Subject. And this on two levels: on the level of action imagined by the Subject, where *X* is the capacity of a supposedly spaceless, timeless, energyless simulacrum of ourselves, in the form of the Agent, to go on iterating "for-ever," and the replacement, the de-written X, is arithmetical passage to the limit $; and on the level of the intelligible repetition, where a cognitive fade-out bounded by @ becomes the de-written replace-ment for the infinite prolongation of meaning sanctioned by the clas-sical conflation of mechanical manipulation of signifiers and cogniz-able manipulation of signs.[50]

As already observed neither of these replacements should be seen as an argued "refutation" of classical infinitism (indeed, it is diffi-cult to imagine how this could come about) nor as a blanket "repudi-ation" of the results—the corpus of classically formulated mathemat-ics—that are the fruits of this infinitism. Rather, in each case, the issue is the kind of persuasion that is to be mathematically acceptable. On the level of the Agent, this entailed an appeal to material, this-uni-verse necessity, and the detailed opening up of the demand for an "idealized" embodied presence to be *actualizable* for it to impinge persuasively on a sign-using mathematician. On the level of the Subject, the move to a limit rested not on any de jure argument about the boundaries or preconditions for rational thought, but on an expe-rientially framed, de facto appeal to the evident unintelligibility of repeating certain cognitive acts (although even without this appeal to experience it would follow that the Subject no less than the Agent could not go on repeating "forever").

It might be objected at this point that a methodological principle that restricted discourse—shackling what could be known, meaning-fully stated, discussed, and imagined, and introducing limits where there were none before—is in principle undesirable and creatively irk-some. What, after all, is the point or benefit in being asked to say and think—to understand as it were—*less* than one can already, in being urged to *not* be persuaded by or understand the classical infinite? To see what is misconceived about such an objection, in terms not manu-factured by the present discussion, it is enough to glance at the origin of modern physics.

Present-day physics is certainly no stranger to the central impor-tance that has to be accorded to an observing, measuring, or knowing subject, a situation that has arisen precisely in relation to the presence of a putative infinity or some "objectively" or "absolutely" given signified. Indeed, the entire intellectual framework of modern physics is founded on two moments—the emergence of relativity theory and of quantum theory—at which something that was previously believed to be absolute, and perfectly well "understood" as such, was denied its purely objective, extratheoretical status and, as a consequence, a subject/observer was introduced into physics in a role crucial to the whole discipline's understanding of itself. Thus, as is well known, relativity theory denied sense to the Newtonian notion of absolute simultaneity of events (which seems to require information to travel at infinite speed) and, by insisting that the speed of information flow be limited to the *constant* speed of light, made it necessary to introduce an observing subject into its formalism and to rewrite the laws of physics in terms of subject-invariant frames of reference. Likewise, quantum theory founded itself directly on the denial of the classical assumption that energy occurred in infinitely subdividable amounts and indirectly on the overturning of the belief that it was possible in principle to measure the state of a system with absolute accuracy. As a consequence physics was obliged to introduce a subject whose observational presence, though it does not produce what is observed, is nevertheless—in some way that remains contentiously theorized and as yet unsatisfactorily explained—held to be inseparable from it.

What is striking here is not the fact of the denials nor their limited nature—they hardly eliminate objectivist thinking from present-day physics—but the ways in which the subject/observers associated with them become part of new ways of signifying that make available conceptual moves and theorizations not accessible within pre-relativistic and pre-quantum physics. Let us return to the principle of epistemic foreclosure. What appears from objectivism's standpoint as the imposition of an unwelcome barrier or limitation—for physics, the constant speed of light, the imprecision of any system; for mathematics, the denial of endless counting—is, from the post-objectivist standpoint, a creative epistemological and discursive principle. The insistence that we cannot know or cognize something becomes productive of new scientific knowledge and new forms of mathematical signifying.

But both these founding moves, the simultaneous exclusions of an absolute signified and inclusions of a subject/observer that originate relativity and quantum theory respectively, have been instituted by a physics riding on the back of classical mathematics. On the back, that is, of a formalism that has patently never rid itself of its attachment to the transcendental and absolute status—the God-given "naturalness"—of the integers and the ad infinitum principle of endless counting that supports it. Thus the attempt by modern physics to de-transcendentalize itself, in the sense of eliminating any appeal to the kind of objectivist metaphysics that is put into play by the classical notions of absolute space and absolute time, cannot but be partial and inadequately attained. Physics can achieve no more than half-freedom from metaphysics, with one foot remaining unnoticed inside the classical paradigm it had thought to have left behind, so long as the absolute conception of number inherent in classical infinitistic mathematics is part of the framework of this attempt. On this one can speculate that the sundry infinities[51] that continue to plague any attempt to put quantum physics on a satisfactory theoretical basis, far from being insignificant "side-effects of the mathematics," minor problems to be ignored or renormalized out of existence, are symptoms—deep-lying and intrinsic—of the very incapacity of "the" mathematics to deliver the kind of nonabsolutist, nonmetaphysical basis that physicists appear to be seeking. An immediate question is whether the realizable mathematics engendered here—by an epistemic move in many ways parallel to that behind the emergence of nonclassical physics—would, if it replaced its classical counterpart, be able to provide such a theorization. More work developing this mathematics is needed before such a question could be answered, but two comments seem pertinent. First, as we saw earlier, the non-Euclidean structure of realizable arithmetic means that any use of it to measure physical observables such as length, time, energy, and so on, would result in a necessary "quantization" of these parameters and the abolition of the idea of an arbitrarily close approximation to a dimensionless "point." Second, whatever its fate, it is certainly the case that, as a noninfinitistic formalism, realizable mathematics would, if it did nothing more, avoid the blatant and unjustifiable mathematical ruses and ad hoc tricks with infinity to which physicists have been forced to resort.

Finally, and not directly connected to the instrumentality of mathematics and its relation to physics, the simultaneous excision of infinity and inclusion of the mathematical Subject lead to the question of how we are to think of the whole mathematical enterprise. Exhibiting mathematics as language and putting the Subject—the historically contingent, culturally produced, intentionally structured, always embodied, sign-creating and -interpreting agency—into the center of our picture make it possible to overturn the whole ideological and rhetorical pretense of mathematical objectivism. This objectivism, a Subject-denying metaphysics locked into place at the end of the last century by Frege and dominating twentieth-century philosophical understanding of the discipline ever since, is almost as old as mathematics itself. The dream, inscribed in the myriad forms that constitute mathematical practice from the moment that this practice is interpreted in classical infinitistic terms, is the fantasy of the absolute: absolute sameness, identity, equality, and permanence of mathematical objects for all times and all conceptual regimes and all historical materializations. In this dream the theorems of mathematics are eternal, transcendental truths written in the timeless present tense of absolute being; and the processes of mathematics—counting, defining, symbolizing—are without limit or end regardless of semiotic agency, human cognitive wherewithal, physical presence, and the exigencies of a never-absent empiricity.

No doubt the idealized imaginings of mathematics answer, as a familiar, unproblematic, innocuous[52] part of everyday wishing and thinking, to the desire for order, regularity, repeatability, form, pattern, and harmony. But poised behind such desires is an absolute desire, introduced into the meaning of number and so into these imaginings themselves, by the presence of infinity. The desire is for no less than that the grandeur and imprimatur of eternity be stamped on the objects of mathematics and the truths one discovers. In this way one can identify with a transcendent being, can move, outside history, in His dominion, glimpse infinity through His eyes, and know about it at least *some* of the truths He knows—for only the "Divine intellect . . . knows all."

The fantasy of a transcendental origin, an ultimate guarantor of Truth unsituated in time, space, or history, for whom or out of whom the infinity of numbers is/was/will be always there, has proved difficult to resist. It leads, as Hilbert knew, into the endless heaven—"paradise"—of Cantor's infinities, and it is, as we have seen, a part of contemporary mathematics and physics. Significantly, Einstein never gave up his adherence to a version of mathematics' transcendental origin despite the fact that his own work led to an overthrow of the determinism that he attached to that origin (hence his much quoted remark about God and dice); and Gödel, though his own metalogical contributions made possible the independence results which have undermined the idea of absolute, set-theoretical Truth, refused to relinquish his hold on the Platonic metaphysics that sustains this Truth. Undoubtedly, the nature of the transcendental origin, the socioculturally articulated identity of what Galileo, Descartes, Newton, Kant, Kronecker, Cantor, Husserl, Einstein, Brouwer, Gödel, . . . called God, varies across persons and places, but the dream of absolutism, independent of these differences, goes back to the beginning. It is Plato's, Aristotle's, and Euclid's dream. It is the Western metaphysical dream of divine and timeless reason and it has continued, transformed but unbroken, passed on in a line from Parmenides' One through the God of Galileo to contemporary science to become the culminating, unstated theism—implicit and unacknowledged—of twentieth-century mathematical infinitism and the technoscientific philosophy of realism that relies so unquestioningly on it.

Compared to the magnificence of such God-guaranteed thought, the alternatives might seem severe: denying mathematics a transcendental origin—rejecting the classical disembodied ghost—makes it impossible for us, as I have argued, to think notions like "endless," "infinite," "forever," "without cease" in relation to counting and number. What before was the smooth-running mathematics of the infinite becomes replaced by the unfathomable complexity of a rewritten finite. But with the disappearance of the "infinite" as such, the idea of the "finite" is itself transmogrified: what is "finite" dissolves into what is or what can be brought into being; it becomes a part, a precondition, a consequence of the very idea of empiricity itself—an empiricity that is no longer the falling short from a divinely apprehendable "ideality," but the fragile, contingent, and disaster-prone condition of all human becoming and all signification in which this becoming takes place. And with this transmogrified finitude comes a transmogrified truth: the whole apparatus of absolute and eternal Truth about numbers falls from the infinite heaven of Plato and the endlessly deferred telos of Aristotle's counting into the entanglements and contingencies of history. Undeniably, seen outside time through the eye of God, the numbers are indeed alien and inhuman. If mathematics has to be spoken of in the vocabulary of "truth," then this can only be truth as a human artifact, truth as a description and production of the one who proclaims it, truth in the sense of Vico's insistence that "the rule and criterion of truth is to have made it" (quoted in Berlin 1976: 20).

But, in fact, as one knows, the desire for the transcendental and its answering fantasy of mathematical absolutism have not been repudiated, resisted, or denied credence. It still holds mathematicians, and hence the rest of us, within its horizon, offering a paradisal theme-park of endless counting and an infinity of infinities without end—a play of pure unattached/disengaged signifiers within a discourse that achieves its sense, reference, and presence by severing any connection to any actual, potential, or imaginable corporeality. The dream of infinity, then, speaks to an old and familiar deity, floating free above the earth-bound, embodied, "finite," but always unfinished business of human becoming.

Do we want to go on dreaming it? If the analysis of mathematics here has any force, if the model it urges is coherent, if its formulation of "number" is intelligible, if it persuades that the dream of endless counting *is* a dream, then the question is rather: *Can* we go on dreaming it—in the same old way?

Appendix

Pre-Elements of Non-Euclidean Arithmetic

Our purpose here is to sketch the character of "number" and its arithmetic set in train by the two limitative principles of realizability and cognizability explicated in the text. The starting point for any examination of arithmetic and number is the activity of counting, that is, the production of signs through the process of repeatedly concatenating instances of some arbitrary but fixed signifier. Thus the primary object of arithmetic study must always be a sequence of iterates such as

0, 00, 000, 0000, 00000, 000000, 0000000, . . .

or

1, 11, 111, 1111, 11111, 111111, 1111111, . . .

or, more formally, through the introduction of the successor function S, the sequence

0, $S(0)$, $S(S(0))$, $S(S(S(0)))$, $S(S(S(S(0))))$, . . .

where the sign ". . ." is to be interpreted as the instruction to the Agent to repeat iterating, insofar as this is possible within the constraints imposed by the condition of realizability. We shall think of the progression of iterates as being bounded by or as tending to a limit denoted by \$, and shall write $i <$ \$ for any iterate i. No assumptions will be made about \$ other than the fact that it exists as the upper bound to the iterates. Of course, \$ is not itself an iterate. Using the ideogram introduced earlier we shall write our primary object of study as

0, 1, 2, 3, . . . \$

and call this sequence the progression of realizable iterates, or simply the $-sequence.

Before we start, let us consider what the orthodox response can be to the idea that the business of counting is subject to some limit. Classically the ordinals form a progression that goes on without end, and issues of entropy, resource limitation, and depletion as they affect the Agent and the consequent fade-out or dissipation in the counting process are devoid of meaning. From such a viewpoint, the only sense attributable to the $-sequence is to interpret it as an initial segment of the endless progression of "numbers" and to identify $ with some number N within this progression. Such a "finitary" identification, what I termed in the text the Columbus, or sudden-death, model, though ultimately at odds with the interpretation of $ to be presented here, has a certain obvious merit, in that it provides an immediate and easy-to-understand situation in which a resource limitation operates— one which, by simply stipulating a largest ordinal, cuts right through the more complex discussion of realizability. Moreover, insofar as it resembles the actual situation of existing computing devices, in which one iterates the operation of adding 1 until the machine gives an error message, it can offer itself as a plausibly pragmatic basis for the development of certain kinds of finitistic arithmetic.

Unfortunately, such a model incorporates the very features that sustain the familiar Euclidean pictures of counting in which the numbers are produced ad infinitum: one proceeds homogeneously in identically repeated steps, except that in this case one falls—without any warning—off the edge at N. As a description of a process of dissipation or fade-out that "finally" occurs, this is artificial, since any such dissipation grows out of the prior execution of the process itself. This means that the emergence of a bound to the process of iteration for *internal* reasons of realizability cannot be put in evidence. And the effect of this bound, the manner in which the iterates tend to $, which conditions their structure *from the outset*, is lost, for all possible different progressions to the limit are projected onto the same initial segment of linear steps determined by N.

In other words, once the inherently non-linear phenomenon of dissipation is recognized as a feature of counting no less than of any other realizable process, it becomes impossible to identify the whole progression of iterates with a homogeneously subdivided line. But clearly the issue cannot rest there, since even if, as we maintain, the iterates are ultimately—as a whole—to be conceived in non-Euclidean terms, their familiar concrete development in everyday counting seems entirely Euclidean. What one wants to say, in fact, is that the iterates are *locally* Euclidean, but to try and elaborate what this means at the present stage of the argument, before the structure of the $-sequence has been examined, would be confusing. For the moment, then, I shall treat the classically inspired zero-dissipation model obtained by identifying $ with N, which I shall call the *Euclidean Projection*, as a useful point of comparison to the realizable iterates, leaving any further explanation of its status until later

Let us start by considering one of the simplest tasks that can be assigned to an Agent in the course of a thought experiment, namely: counting from 0 to some iterate k in the $-sequence and then returning to 0 with the information about k the Agent had been dispatched to establish. To concretize this kind of primary thought experiment we might think in terms of the energy requirements of the Agent: first, the energy $E_1(k)$ to count up to k, then the energy $E_2(k)$ to perform the relevant computational task on k, and finally the energy $E_3(k)$ required to complete the circuit and return to 0 with the information provided by the computation. In exactly similar manner we might think also in terms of the requirements of space, the requirements of time, and the effects of entropy, in each case introducing three functions of k more or less parallel to E_1, E_2, E_3. The resulting formalism, though it would put the working of realizability into a concrete and realistic setting, would be unwieldy and would mask the underlying ground rules for realizable iteration that I wish to reveal here. In view of this, certain simplifying assumptions will be made throughout the following. Specifically, I shall ignore E_2, put both E_1 and E_3 as equal to the identity function, and speak in general terms of the effects of increasing resource depletion or dissipation. The effect of these assumptions will be to imagine the Agent as counting up to k and having to return by counting down from k.

On this basis we observe that for small enough k the Agent will be able to complete a circuit and return to 0. As k is increased the effects of dissipation and resource depletion make the completion of the circuit increasingly difficult. For large enough k no such completed circuit can occur, since the journey up to k will have depleted resources to such an extent that only a return by the Agent to some iterate larger than 0 will be possible. In relation to such circuits there are, then, three possibilities for any given iterate: k is definitely "small enough" for a return to 0; k is definitely "too large" for such a return; k is neither small enough nor too large and no definite decision is taken about whether a return to 0 from k occurs. We introduce a cut or point of no return for the Agent, which we denote by a_0, and write $k < a_0$ for the first possibility and $k > a_0$ for the second possibility. Note that a_0 is not itself conceived as an iterate: indeed, the nature of the indeterminacy which occurs at a_0 will be a function of the whole $-sequence, of the way the elements of this progression tend to or are bounded by the limit $. (In terms of the Euclidean Projection, where $ is identified with N, the interpretation of a_0 is simply the midpoint of the sequence from 0 to N.)

Observe that the operation which creates a_0 from the whole $-sequence as a point of no return to 0 can be repeated: for a counting to k for large enough k there will be a point, call it a_{00}, beyond which no return to a_0 will be possible. And beyond this one can define a yet higher point of no return to a_{00}, and so on, dividing up the region beyond a_0 into ever more remote zones. However, such zones do not play a crucial role in the outline of realizable arithmetic being sketched here, and we shall not pursue their significance.

To say more about a_0 we—that is, the mathematical Subject—must introduce a new level of notation. We define the binary operation $+$ of addition as the concatenation of two iterates, so that $j + k$ is to mean the iterate obtained by counting to j and then counting a further k steps. Obviously the sequence of iterates is not closed under addition: for large enough k the effect of resource depletion will mean that $k + k$ is undefined—there being no iterate that can correspond to it. Thus a_0 is the *exit point* for addition of realizable iterates in the sense that:

(1) $k + j < \$$ for $k,j < a_0$; and $k + j$ undefined for $k,j > a_0$

Observe that as soon as we consider repetitions of $+$, then there will be exit points that occur before a_0. Thus by the same depletion consideration as above there will be $k < a_0$ such that $k + k < \$$ but $k + k + k$ will be undefined, giving rise to an exit point a_1 for 2-fold addition of iterates satisfying

(2) $h + j + k < \$$ for $h, j,k < a_1$; and $h + j + k$ undefined for $h,j,k > a_1$

Likewise, there will be exit points a_2, a_3, for 3-fold and 4-fold addition, and more generally for $i > 0$, a_{i-1} for i-fold addition $j_0 + \ldots + j_i$ of iterates, provided $i < \max (j_k) < a_i$.

Let us denote by b_0 the limit of this downward sequence $a_0 > a_1 > a_2 > \ldots$ of exit points. To interpret b_0 we need to introduce a new level of notation: specifically, the binary operation $\times$ of multiplication as repeated addition, by considering $k \times j$ as the iterate in the $\$$-sequence obtained by adding j to itself k times. Clearly b_0 is the exit point for multiplication in complete analogy to a_0 for addition, so that we have

(3) $j \times k < \$$ for $j,k < b_0$; and $j \times k$ undefined for $j,k > b_0$

This whole procedure can be repeated. As before there will be a downward sequence from b of exit points, $b_1 > b_2 > \ldots$ for 2-fold, 3-fold, $\ldots$ multiplication. And, again as before, we can denote by c_0 the limit of these points, and interpret c_0 by introducing a binary operation, exp, of exponentiation, with the meaning of repeated multiplication. Observe that c_0 is the exit point for exponentiation:

(4) $j(exp)k$, $k(exp)j < \$$ for $j,k < c_0$; and $j(exp)k$, $k(exp)j$ undefined for $j,k > c_0$

We could now form a downward sequence from c_0 of exit points for repeated exponentiation whose limit would be d_0; and so on. It would seem that the idea is clear enough: one could go on repeating the procedure, forming new descending sequences, which determine new cuts in the $\$$-sequence marking new exit points for hyperexponentiation as repeated exponentiation, hyperhyperexponentiation as repeated hyperexponentiation, and so on.

In terms of the Euclidean Projection: $\$ = N$ and a_0, b_0, c_0, d_0, and so on, correspond respectively to solutions of the equations:

(5) $k + k = N$, $k \times k = N$, $k(exp)k = N$, $k(hyperexp)k = N$, . . .

Thus, if $N = 10^{40}$ for example, then $a_0 = 5.10^{39}$, $b_0 = 10^{20}$, $27 < c_0 < 28$, $2 < d_0 < 3$, and further repetition will result in a descending infinite sequence of real number solutions to the corresponding equations in (5).

An obvious question arises: what, given our interpretation of the iterates in terms of realizability rather than their images in the Euclidean Projection, would be the meaning of such a sequence of descending solutions? In order to make sense of this question, let us return to the idea of indefinability as it occurs in the picture of an exit point. Such points bear witness to the occurrence of a spillover: the realizability condition entails that simple operations on iterates— conventionally defined addition, multiplication, and so on—cannot always be assumed to produce an iterate. We might nonetheless try to give meanings to the undefined results of these operations, and thereby include such spilled over items as "numbers" within an enlarged arithmetic.

The simplest operation is +. We saw that provided j and k are below a_0, $j + k$ is defined; meaning that we are able to identify it with some iterate $h < \$$. But how are we to assign a meaning to $j + k$ for $j,k > a_0$? Clearly, whatever significance we attach to such a sum, the entity cannot be an iterate, it cannot represent a position within the sequence of realizable iterates that an Agent could reach by counting from 0. And if we want to think of it as attached to a plurality in some sense, then the members of any such plurality cannot be countable; their "number" must be brought into being in a single act of stipulation and would appear, then, as a pure cardinal p with the property that $\$ < p$, that is, $i < p$ for all iterates i. Thus, if we wish to define the operation of addition over the whole sequence of realizable iterates, and not just to those below a_0, we must imagine a region of "numbers," call them *transiterates*, situated beyond $\$$.

To this end, denote by $\$1(+)$ the set of all ordered pairs (j,k) for iterates j,k, where two pairs, (i,h) and (j,k) are identified if the rules of iterate addition entail the equality $i + h = j + k$. Thus, for example, from the rule $1 + n = n + 1$ for $n + 1 < \$$, we have $j + k + 1 = j + 1 + k$, so that $(j,k+1)$ is identified with $(j,1+k)$ provided $k + 1 < \$$, and so on. We interpret $\$1(+)$ as the collection of all 1-fold sums, $j + k$, of iterates j, k.

Closing the iterates with respect to $+$ in this way applies only to 1-fold addition and says nothing about repeated application of the operation. More generally, for an iterate i, we can repeat the above procedure and denote by $\$i(+)$ the collection of all permissible i-fold sums $j_1 + \ldots + j_{i+1}$ of iterates j_y. This process of augmentation can in turn be repeated and then combined to produce a region, which we denote by $\$(+)$, of transiterates that can be considered as the $\$$-closure of the iterates with respect to addition.

In an exactly parallel way the whole process can be carried out for $\times$ to produce further transiterative enlargements: first $\$1(\times)$, then $\$i(\times)$ and then $\$(\times)$ as the $\$$-closure of the iterates with respect to $\times$. Likewise for the operation *exp* of exponentiation to produce $\$(exp)$ as the $\$$-closure of the iterates with respect to *exp*; and then again for the operation *hyperexp* to produce $\$(hyperexp)$, and so on, for further arithmetic operations.

Observe that not only can the individual transiterates not be the result of counting from zero, but the region $\$(+)$ of transiterates, and likewise $\$(\times)$, $\$(exp)$, and so on, cannot be brought into existence as an $\$$-sequence, but must be posited by a process that transcends any iterative counting from 0. This follows from the fact that for any given i, there exists $j > i$ such that for some z in $\$j(+)$ we have $z > y$ for all y in $\$i(+)$. This means that any counting of $\$(+)$ would have to enumerate the whole of $\$i(+)$ before it could count a final segment of $\$j(+)$; if we repeat the process on $\$j(+)$, and so on, there will be a $k < \$$ such that no z in $\$k(+)$ can be counted. The transiterates, in other words, insofar as they are "numbers," must be considered not as ordinals but as "uncountable cardinals." If the elements of $\$(+)$ cannot be ordered into an $\$$-sequence, a natural question to ask is whether it makes sense to insist that they be totally ordered; and the same question arises for all subsequent transiterative extensions, namely $(\times)$, (exp), and so on.[53]

Returning to the issue of closure, we can ask about the closure properties $(+)$ itself enjoys as the result of the above construction. By definition, for a,b in $(+)$ there will be $k,j < \$$ such that a is a k-fold sum and b is a j-fold sum of iterates. Clearly, if $k,j < a$ then $k + j = i < \$$, and so $a + b$, as an i-fold sum, will belong to $(+)$; conversely, if $\$ > k,j > a$, then $k + j > \$$ and there will be no entity in $(+)$ corresponding to $a + b$. Thus a sufficient condition for $+$, in 1-fold form, to be defined for transiterates is that the summands are "small enough"—that is, bounded above by a limit which we shall denote by $a^\$$. In other words,

(6) $a + b$ is in $(+)$ if $a,b < a^\$$

The question of the closure of $(+)$ with respect to 1-fold addition can be repeated for 1-fold multiplication, 1-fold exponentiation, 1-fold hyperexponentiation, and so on, to yield exit zones, $b^\$$, $c^\$$, $d^\$$, and so on, for these operations on $(+)$, satisfying the natural parallels to (6).

Let us return to the more general sense of the "closure" of realizable arithmetic. Writing the successor operation S as O_0, addition as O_1, multiplication as O_2, exponentiation as O_3, hyperexponentiation as O_4, and so on, one forms the sequence of operations

$$O_0, O_1, O_2, O_3, O_4, \ldots @$$

where $@$ is the cognitive limit of the Subject identified earlier. The "closure" brought into play here can be seen more readily if we simplify matters by considering the unary rather than the binary version of the arithmetical operations. Thus, instead of $i + j$—that is, $O_1(i,j)$—let us consider $i + i = O_1(i,i)$ which we shall write as a unary function $U_1(i)$ of i, and similarly with the other operations, so that $i \times i = U_2(i)$, $i^i = U_3(i)$, $i(hyperexp)i = U_4(i)$, and so on. The exit point for unary addition is the limit of the solution to the equation $z + z = k$ as k tends to $\$$. More generally, for any given n the exit point for U_n will be given by the limit as k goes to $\$$ of the solution to the equation $U_n(z) = k$. If the exit point of U_n is less than the iterate j, then $U_n(j)$ is undefined and so therefore is $U_n(j)$ for all $j > i$. Now apart from unary addition which has the value $U_1(1) = 1 + 1 = 2$ for the argument 1, all the subsequent operations have the property that $U_n(1) = 1$, so that asking whether any U_n is meaningful for the iterates is equivalent to asking whether $U_n(2)$ is defined, that is, whether the exit point for U_n is greater than 2. Lacking any external determination of \$ and @ we can say nothing more specific than this.

However, to get the feel of the restrictions imposed by the limits here, we can map the situation onto an initial segment of the classical integers and interpret the arithmetical operations accordingly. Thus, substituting, say, $N = 10^{80}$ for \$ and calculating values for the argument $z = 2$ we have $U_1(2) = 4$, $U_2(2) = 4$, $U_3(2) = 16$, $U_4(2) = 2^{16} = 65536$, so that $U_5(2) = 2^{65536} > 10^{80}$ making U_5, the hyperhyperexponentiation, undefined. But the choice of 10^{80} as a "value" for \$ here is purely arbitrary, and had we taken instead $N = 10^m$ for $m > 10^5$, then U_5 would have been defined. Indeed, as is obvious, for any given integer k there will be a value of N such that with respect to this N we have U_k defined. But it is precisely this ad infinitum ability to extend N as far as one pleases that makes the resort to a classical model—whatever explanatory worth it has—ultimately misleading. For, after all, it is in the process of questioning the infinitistic extendability of N that the whole notion of realizability makes its appearance.

The conclusion to be drawn here is that the outer range of trans-iterates that "exists," that is available to the arithmetical Subject—which fixes the closure of the model of arithmetic being sketched here—does not arise from an externally imposed numerical limit. Rather the reverse is the case: what counts as "number" is determined by the cognitive limits that operate on an embodied Subject called upon to manipulate—read, write, process, handle, decipher—materially presented signs created by the introduction of new notations. Thus, for example, consider the operation U_5 of unary hyperhyperexponentiation. Starting with $U_5(1)=1$, the next step, $U_5(2)$, yields an expression in which 2 is raised to a stack of 16 exponents. Manipulating such expressions and relating the results to other number expressions presents obvious difficulties—difficulties which increase cumulatively as further operations such as U_6, or perhaps U_7, are considered, until they become, as I have argued, cognitively insurmountable.

Let us summarize the meaning of "number" being suggested here as a realizable and cognizable replacement of the classical progression of natural numbers. By *number*, then, is meant the sequence 0, 1, 2, 3, . . . \$ of iterates, ordinals, that can be counted into being by an Agent constrained by realizability together with those transiterates (uncountable cardinals) that can be countenanced by a cognitively constrained Subject through the introduction of arithmetical operations and the notations from which, for such a Subject, they are inseparable. Writing INTEGER for the extension of "number," ITERATE for the sequence of iterates, and TRANSITERATE for the remainder at any given stage, we have

INTEGER = ITERATE + TRANSITERATE

where + denotes the process of augmentation described above.[54] If we want the maximal form of this process of augmentation—the closure of the model—then we have

TRANSITERATE = \$(+) + \$(×) + \$(exp) + . . .

where ". . ." means that the introduction of the arithmetical operations is to be continued for as long as it is feasible to do so. More briefly, denoting by ++ repeated augmentation:

INTEGER = ITERATE ++ \$(0_i) for $i <$ @

To fill out the concept of "number," and give a further feel to the consequences of the limits @ and \$, we need to describe the structure of the realizable rationals. We shall follow the orthodox construction of the rationals from the natural numbers. This exhibits them as equivalence classes of ordered pairs of integers with respect to the equivalence relation *eq* given by:

(7) (a,b) *eq* (c,d) if and only if $ad = bc$

The class of all pairs equivalent to (a,b) is denoted by a/b and the whole arithmetic of the rational numbers stems from the definitions:

(8) $a/b + c/d = (ad+bc)/bd$

(9) $(a/b)(c/d) = ac/bd$

The natural analogue to ratios of natural numbers would be ratios of elements of ITERATE, that is a/b for $a,b < \$$, and we would like to mimic the classical construction given by (7), (8), (9) just for this case. But it is more convenient to apply the construction to a/b for all numbers a,b that are defined as belonging to INTEGER, and then separate out the ones that interest us: to this end we shall use the term *rationals* (extension RATIONAL) to denote ratios a/b for $a,b < \$$, and call all other ratios *transrationals* (extension TRANSRATIONAL).

Let a, b, c, d be iterates. For the addition of rationals the right-hand side of (8) gives rise to three cases:

(10) if $ad, bc < a_0$ and $b,d < b_0$, then $ad + bc < \$$ and $bd < \$$ so that $a/b + c/d$ will be a rational

(11) if $a^\$ > ad, bc$, then $a/b + c/d$ will be a transrational

(12) if $ad, bc > a^\$$, then $a/b + c/d$ will be undefined

where in (11) and (12) the bound $a^\$$ is the exit point for $+$ on $\$(+)$, given in (6).

Note that for (11) to be valid we are supposing that the products ad, bc, bd are defined. To ensure this we have to assume that

(13) INTEGER contains the transiterates $\$1(\times)$ representing 1-fold multiplication

A sufficient condition for this to be so is:

(14) INTEGER contains the transiterates $\$(+)$ representing all repeated sums

If we restrict attention to rationals x/y satisfying $x/y < 1$, then $a < b$, $c < d$ and so $ad, bc < bd$ in the above, and we can write (10) and (11) in simpler but weaker versions:

(10a) if $bd < a$ and $b,d < b_0$, then $a/b + c/d$ is rational

(11a) if $bd < a^\$$, then $a/b + c/d$ is defined as a transrational

In particular, the sum

(15) $1/a + 1/b = (a + b)/ab$ is always defined

If we consider k-fold sums of rationals for $k < \$$ the same analysis applies but the assumption (13) has to be strengthened: to ensure definability we need to have introduced transiterates $\$(\times)$ representing k-fold products of iterates for all $k < \$$.

For multiplication of rationals the situation is much simpler. According to (9), for iterates a, b, c, d we have immediately

(16) $(a/b) (c/d) = ab/cd$ is always defined

provided of course that (13) holds. And, in particular,

(17) $(a/b) (c/d)$ is rational when $a,b,c,d < b_0$

If we wish to consider repeated products of rationals, then we can ensure definability only if we modify assumption (13) to include all transiterates in the region $\$1(exp)$, or modify (14) to include $\$(\times)$.

To get an idea of how our ratios are distributed we can compare them with the classical rationals. As one knows, the classical rationals are dense in themselves—between any two there is always a third. Does this property hold for the rationals and transrationals introduced here? Specifically, for given $a/b < c/d$ does the following hold?

(18) There exist iterates e,f such that $a/b < e/f < c/d$.

For a/b and c/d rational, consider the special case of this when $a = c = 1$, $b = d + 1$:

(19) There exists e/f such that $1/(d + 1) < e/f < 1/d$

Clearly (19) is equivalent to

(20) $de < f < de + e$

Obviously e cannot be equal to 1. If $e > 1$, then $f > d + d$. But for $d > a_0$ this means that $f > \$$, and so e/f cannot be rational. Thus there exist pairs of adjacent rationals, in the sense that there exists no rational between them. If, however, we drop the requirement that e/f be rational, then, assuming (14) to hold, (19) always has a solution for rational a/b and c/d, since we can put $e = 2$ and $f = 2d + 1$, where

$f = d + d + 1$ is a transiterate in $2(+)$. More generally, (18) holds for e/f transrational provided, again, that (14) is assumed to hold.

Thus, the transrationals are dense in the rationals which are not dense in themselves. Similar arguments show that the transrationals cannot be dense in themselves, that is, there exist transrationals a/b and c/d such that (18) has no solution. To see this, consider (19) in the case that $a > a^\$$. It follows from (20) that $f > a + a > t$ for all t in $\$(+)$. For f to be defined a further extension of the transiterates beyond that given by (14) is required. But then the argument could be repeated for an element a of this extension, and so on, until no more extensions are available.

The distribution of the rationals is thus one of increasing sparseness as they become smaller, reducing ultimately to sequences of the form $1/i$, $1/i+1$, $1/i+2$, . . . once i exceeds a_0. Likewise, as the denominators of the transrationals increase, so the transrationals themselves thin out and eventually reduce to sequences of the form $1/b$, $1/c$, . . . where b, c, . . . will depend on the value of @, that is, on the regions of transiterates one can assume to be present in INTEGER.

Finally, note that in the classical development of number one would at this point, having defined the whole numbers and rationals, introduce the real numbers as Dedekind cuts or Cauchy sequences in the field of rationals. A similar procedure would work here—though the resulting continuum would obviously look very different from the classical one. It would, however, be premature to pursue this in advance of a more thorough and systematic construction of the realizable numbers than given in this brief presentation.

Reference Matter

1. The conflict about language cuts very deep—deeper even than the much-publicized "linguistic turn" attested to by analytic philosophers. In this turn every major area of their subject has defined itself in relation to linguistic issues such as representation, sense, reference, speech acts, translatability, and meaning that have dominated the analytic tradition since Gottlob Frege's work on formal mathematical systems. For the philosopher Nelson Goodman it has its origins in the Enlightenment when "Kant exchanged the structure of the world for the structure of the mind, continued when C. L. Lewis exchanged the structure of the mind for the structure of concepts, and now proceeds to exchange the structure of concepts for the structure of the several symbol systems of the sciences, philosophy, the arts, perception, and everyday discourse" (1978: x). More radically, Richard Rorty (1979), urging that the problem posed by language for analytic philosophy is more extreme and has the character of a crisis, denies that a solution can be found within a Kantian-inspired framework, however extended or modified; and he argues for a new approach, a pragmatism after John Dewey that looks to Heidegger's and the later Wittgenstein's more revolutionary writings on language. Outside philosophy and the other disciplines listed there is also semiotics (of which much more below); and there is psychoanalysis: the rupture in the notion of rational discourse brought about by Freud's theorization of the unconscious, and Jacques Lacan's reading of Freud, in which the unconscious is to be seen as "structured like a language."

2. Consider the titles of works that span the mathematical literature from the professional and technical—e.g., *The Grammar of Science*, Karl Pearson's (1937) survey of statistics early in this century—to the popular—e.g., *Number: The Language of Science*, Tobias Dantzig's (1953) widely read history of number notations.

3. I take this nomenclature from Putnam (1983: 272), though the formulation of it given here is due, apparently, to Michael Dummett (1975), who refers to it as "realism."

4. And this is the case whether the apology is on the level of popular comment by one who writes from outside the subject about its wonders or is part of an elaborated thesis about the nature of consciousness and human reasoning by a practicing mathematician/physicist. For the former, see Martin Gardner: "How mathematicians who pretend that mathematical structure is not 'out there,' independent of human minds, can view successive enlargements of the Mandelbroit set and preserve their cultural solipsism is hard to comprehend" (1989: 26; see also Gardner 1983 for many similar and more extended versions of this sentiment). For the latter, see Roger Penrose: "I strongly take issue with the assertion that Gödel's argument does not directly support Platonism. To deny mathematical Platonism in any form is to deny that the truths of (say) arithmetic have an absolute validity independent of human culture. . . . Like any other convincing mathematical argument, such reasoning [Gödel's theorem] helps to persuade us of the truth or falsehood that is evidently not up to us to decide but merely to discern" (1982: 639; see also Penrose 1989 for a general elaboration and defense of this comment). For a patient and rigorous analysis of how Platonism manages to seal itself off from any form of specifically *constructivist* criticism and protect its adherents by a series of self-confirming moves, see Stolzenberg 1978.

5. Thus, see for example such works about mathematics as differently motivated and organized in relation to their themes and philosophical concerns as Steiner's *Mathematical Knowledge* (1975), basically a scholastic reaffirmation of status-quo Platonism, and Kitcher's *The Nature of Mathematical Knowledge* (1984), an attempt to reinstate a version of J. S. Mill's empirically based account of mathematics. What these and numerous similarly titled works offer is not one view among others—mathematics as knowledge set against a conception of it as language or writing or instrumental reason or discourse or whatever—but *the* (post-Fregean and ultimately Kantian and Cartesian) view of the practice: mathematics as a "realist" science engaged in the production of true and certain knowledge about a preexistent domain of "sets"; the examination and critique of this knowledge, its supposed apriority, its certainty, its foundation, and so on, being—without any need to justify the point—the principle through which the subject is to be perceived and philosophized. Such works, therefore, beg and/or elide in their very titles the very questions—how is mathematical "reality" constituted, how do its assertions "about" this reality come to be "true"—they suppose themselves to be investigating.

6. Of the main current forms of constructivism—Brouwer's intuitionism, A. A. Markov's recursive formalism, Everett Bishop's computationalism—

that of Bishop is the most logically stringent, in the sense that the first two, though logically incompatible with each other, can each be obtained from Bishop's constructivism by the addition of an assumption not acceptable within the latter system. See Bridges and Richman 1987 for a straightforward exposition of the three approaches and Stolzenberg 1970 for a more detailed and searching appreciation of Bishop's underlying philosophy.

7. As the title of this essay indicates, the question of reinstating the body will be all-pervasive and crucially important in what follows. For a differently structured, angled, and focused but in many ways parallel insistence on the need to counter objectivist epistemologies and insert the body back into "knowledge" and the working of the imagination, see Johnson 1987.

8. See Derrida 1976 for a meditation on what "writing" is to mean when it is freed from its subordination, as a secondary phenomenon, to a prior speech and released from the "metaphysics of presence" such secondariness carries with it; and Derrida 1982 for an elaboration of the alphabetic prejudice that supports this metaphysics. For a differently styled, short, and admirably cogent treatment of the alphabetical prejudice—one not embroiled in philosophical issues of "presence" and the like—see Harris 1986.

9. This richness is inseparable, however, from the impossibility of giving Peirce's definition of sign an unambiguous or univocal reading, particularly in relation to Saussure's signifier/signified opposition. Consider, for example, how one is to map Peirce's "interpretant" onto Saussure's vocabulary. Is it to be made to correspond more or less to the signified, to be "virtually synonymous" with it, as Silverman suggests (1981: 15)? Or should it, as Pavis asserts, be seen as another name for the signifier, with Peirce's "representamen" being matched to the signified (1982: 14)? It would not be difficult to find usages and texts by Peirce which justified each of these choices, and one is left with the sense of an irremediable equivocation in Peirce's understanding of a sign quite contrary to his avowed scientific program for semiotics. Given the essential slipperiness of signs and their "ungroundedness"—Saussure's simple opposition is itself ultimately unstable—one should not perhaps be unduly disturbed by this.

10. On the difficulties and obscurities that thread through Peirce's idea of "object" and his insistence that all signs necessarily "refer to" or "represent" or "stand for" their object, see Greenlee (1973: Ch. 3).

11. Thus Derrida's repudiation of the structuralist basis of Saussure's semiotics and his attempt to understand "sign" through the concept of

differance, a many-sided amalgam of difference (static, closed, spatial) and deferring (dynamic, open, temporal). Akin to this is Bourdieu's rejection of "structure" as an explanatory anthropological principle and his dismantling of the primacy of abstract system (language, rules) over the employment and instantiation of it (discourse, everyday practice), with the intention of reinstituting change and process into anthropology.

12. In relation to mathematics two points need to be made about the finite/infinite coupling. (1) There is a perfectly acceptable *intrinsic* definition of finite and hence, by negation, infinite, that seems to avoid the circular dependence of the two notions. Thus, one defines a set to be finite if it cannot be put in one-to-one correspondence with any of its proper subsets, and infinite otherwise. But the whole logical and set-theoretical apparatus with respect to which such a definition makes sense is constructed on the prior assumption and meaningfulness of an "infinite" domain of discourse. So that, however satisfactory for the purposes of classical, infinitistic mathematics, the definition has no value in an investigation, such as the present, of the status of such mathematics. (2) All ideas of the *metaphysical* infinite—essentially those identifying the infinite with the "absolute" or any of its totalized variants—would put the sort of finite/infinite duality concerning us here into question. Thus for Hegel anything which is conditioned, conceptually limited, or bound in some way or other is by definition *finite*. Thus, for example, the integers' being contained within the rationals is an instance of the "finite infinite" or, as he says, the "bad" infinite; likewise the rationals, the reals, and all the other familiar infinite sets of (infinitistic) mathematics are all bad in relation to the "good," absolute totality of all possible objects. Though too crude and extravagantly dismissive to be of much use in discussions of mathematical infinitude, Hegel's bad/good distinction finds an echo both in Cantor's writings on the absolute and in the contemporary distinction drawn in infinitistic set theory between sets and proper classes: both echoes being, on the present account, the inevitable return of an unacknowledged and buried theism.

13. See Hughes and Brecht (1978: 22–23) for this and the other paradoxes of infinity attributed to Zeno, as well as succinct formulations of many other paradoxes, antinomies, and paralogisms. The version I quote in the text is taken by them from Benardete 1964, where it is discussed in some detail.

14. For ease of explication I have overstated Platonist confidence in the finality of the current axiomatic solution to the question of infinity. Even the most deeply blinkered Platonist is hard put to avoid a certain unease

when confronted by the independence results (e.g., the continuum hypothesis or the axiom of choice) in axiomatic set theory. Of course, Platonism, like any long-lasting system of metaphysics, is flexible in the face of contrary evidence, and these results, together with the difficulties they raise, can be negotiated. Thus, Platonists can preserve their belief in Truth and a preexistent realm of timeless objects by asserting that (say) the continuum hypothesis is either true or false in some absolute sense, but that we simply do not *know* which it is, either because of our inherent limitations or because we have not yet formulated the correct assumptions from which to decide the issue. Gödel held to a position somewhat like this.

15. Quoted in Tiles (1989: 29) where the actual/potential distinction is discussed in relation to Aristotle's formulation of the logic of classes. For a sustained polemic refusing the coherence of the idea that the potential infinity can be separated off from the actual, and an imaginative and paradox-filled embrace of the metaphysics of infinity—of, as he puts it, "playing God"—that issues from such a refusal, see Benardete 1964.

16. It is not the transcendental nature of a unitizing ego as such that is the problem. Any attempt to use Kant's writings as a basis for thinking about infinity has to contend with an irreconcilable tension and much consequent obscurity between two Kantian first principles: the absolute infinitude of God and the essential finitude or empiricity of all human knowledge. Thus the question of infinity is begged deep within his characterization of the precondition for rational thought: our very intuitions of space and time—within which geometry and arithmetic are grounded for him—have built into them the possibility of an infinite iterability, the preacceptance, in other words, of a progression ad infinitum of both prolongation and subdivision. It seems accepted and well recognized that mathematics for Kant "requires that there be some source in our cognitive apparatus for some sort of representation of infinity" (Parsons 1983: 109). The existence of such a source, sensed as a moral absolute within us or, what amounts to an aesthetic version of the same, as the means for recognizing the mathematically sublime, was openly acknowledged by Kant: "the *mere ability even to think* the given [i.e., actual] infinite without contradiction, is something that requires the presence in the human mind of a faculty that is itself supersensible" (1986: §26). Against this, Kant accepts that since only objects of experience can ever be *known*, infinity, potential or actual, which manifestly is not encountered in any experiential or empirical situation, can only be an Idea of reason that impinges on us in a purely regulative manner—we act *as if* it were the case that there is an infinite totality or an infinite reality. Plainly, neither the invocation of a felt

interior absolute—God—nor the transposing of the question into the
realm of the "as if" is of much use if it is the very cognitive content, metaphysical implications, and semiotic status of what is being absolutized and
regulated, of what we are to mean by the *supersensible*, that is at issue. For
further (more sympathetic) discussion of the regulative idea of infinity, see
Moore 1990; and for an elaboration of Kant's theism as a justificatory
"source," see Derrida 1981a.

17. I am referring here to the prevailing and near universally accepted
view. There are many individuals who have argued from a multiplicity of
perspectives against the "naturalness" of the infinite progression of integers and for a more critical and rigorous stance towards the meaning of
the whole numbers. Thus Poincaré's (1913) attack on the circularity of the
standard justification for the principle of mathematical induction; Wittgenstein's critical meditations of what it means to follow or not follow the rule
of counting (1978 and Wright 1980); misgivings about the very meaningfulness of large number notations in Van Dantzig (1959) and Bernays
(1964); Yessenin-Volpin's (1970) sketch of an ultra-finitistic account of
mathematics; Isles's (1981) proof-theoretic development of Yessenin-
Volpin's ideas; Lorenzen's (1955) strict operationalism; Rashevskii's (1973)
direct attack on the "dogma" of the natural numbers (compare note 43
below); Parikh's (1971) work on induction and the concept of feasibly
long proofs; Van Bendegem's (1986) finitist approach to applied mathematics (compare note 21 below); Nelson's (1986) critique of arithmetical
impredicativity. In some way or other, all of these argue for or operate
within or arrive at nonclassical frameworks of various sorts that systematically exclude infinity from their proposed treatments of mathematics. In
particular, echoes and prefigurements of some of the ideas in the present
account occur throughout Isles's writing and in the critical protosemiotic
positions adopted in the all too brief papers by Van Dantzig and Rashevskii
cited here.

18. See Smith 1988 for a treatment of this issue—the raising of relativist
specters and nightmares of unrestrained subjectivism by an absolutism
under threat—in the context of denying an objectivist basis for "value"
rather than "number," as well as an attempt to overcome the whole binary stalemate of such a maneuver.

19. One very important philosophical root of the social construction
case for mathematics is Wittgenstein's way of talking about mathematical
activity in terms of "language games" and "forms of [social] life." See
Bloor 1976 for a lucid attempt to give some anthropological bite to
Wittgenstein's remarks about the possibility of "alternative" mathematics
and for a reappraisal of J. S. Mill's much earlier, cruder empiricist approach

to the social construction of mathematical knowledge. Also, in relation to the social nature and construction of objectivity itself, see Bloor and Barnes 1982, Porter 1990, Restivo 1983 and 1990. For a useful analysis and survey of "non-objectivist" philosophies of mathematics in the context of his own brand of social constructivism, see Ernest 1991.

20. In Egyptian orthography a determinative for plurals was introduced around 2500 B.C. Instead of repeating an ideograph "X" to give "XX" for the dual (two X's) and "XXX" for the general plural (many X's), scribes wrote "X. ." and "X. . ." respectively. One might, without being too fanciful about it, understand our concern to open out the ideogram ". . ." as an attempt to explicate the limit of the idea of the plural as such, that is, the idea of "many" in its purest, arithmetically stripped-down form.

21. Thus the system presented by Van Bendegem (1986), which develops a finite model of arithmetic with the aim of providing a foundation for mathematics insofar as it is empirically useful. Van Bendegem's approach, an impressive demonstration of what can be achieved from the very simplest assumption (basically, that the mathematician has at his/her disposal only a single sheet of paper), would be described in terms of the one developed below as a zero-dissipation model of resource depletion whose arithmetic would be a linearized projection of the non-Euclidean version to be constructed here.

22. See Putnam (1983: 284–86) for an idea along these lines—basically an attempt to use the intuitionistic logic of implication applied to "vague" predicates. But the standard mapping of intuitionistic logic onto classical logic preserves chains of implications, so that from the perspective advanced here such a suggestion fails to meet the point about changes that occur solely within the course of a chain of iterated syllogisms. Putnam also cites another, more interesting proposal from Rohit Parikh for a logic or language of vague predicates. Parikh suggests a restriction in the length of what would be allowed as a proof and with this an acceptance of the idea of a logic or language being locally consistent but globally inconsistent. Parikh's suggestion is articulated within the framework of classical, infinitistic logic; if one made the restriction on length follow from an *internally* generated limit and replaced global inconsistency by global indeterminacy, then the picture he offers is akin in some respects to the one I shall offer here.

23. The phrase is Gregory Bateson's. See Bateson 1972 where "the difference that makes a difference"—the definition of *information*—is made into a constitutive principle of an ecologically based model of communication, one whose insistence that all communication is from the start riddled with metacommunication is echoed in the model of mathematical activity

developed here. Bateson's framework has been extended and given a more abstract and systematic treatment by Anthony Wilden; see Wilden 1972 and 1987 for an extended theorization of the semiotic and political dimensions of the analogue/digital opposition central to any communicational analysis of "difference."

24. It should be noted, in view of the subsequent importance of the idea of realizability in the account here, that though Kuhn's idea of a thought experiment is a very wide one, embracing different types of imagined scenarios, including "situations that could not be fully examined and that need not occur in nature at all" (1977: 240), his discussion assumes that these situations are such that they could *in principle* so occur. Were this not the case, much of what he says about the relation between thought experiments and knowledge would be meaningless.

25. For an illuminating account of the use, development, and importance of *Gedankenexperiment* in the natural sciences from the eighteenth century, together with a comparison to fictional experimentation in literature, see Walter Moser's "Experiment and Fiction" in Amrine 1989. The subject of thought experiments is more topical than I realized. Two book-length studies, Brown (1991) and Sorensen (1992), appeared while I was completing my own text.

26. The semiotic model about to be presented here is in certain respects a more developed version of the one given in Rotman 1988. For comparison the reader should note that what I here call the mathematical "Subject" is there called the "Mathematician," and that the revisionary program in relation to infinity toward which the account here is slanted is absent from the earlier version.

27. This rigor and complete lack of ambiguity is a *claim* by the mathematical community. It is included as a characteristic of formal mathematics within the model of mathematics being constructed here on a purely descriptive basis. Indeed, not only is the claim to have achieved objective rigor within an unambiguous formalism not immune from philosophical skepticism—witness, for example, Wittgenstein's critique of following the (unambiguous) rule of counting—it is also seen as relative and contingent as soon as rigor is perceived as an historical achievement and not as a transcultural and objective given.

28. Those familiar with deconstruction will make an immediate connection between Derrida's "logic of supplementarity" and the description in the text of the fate of the Code/metaCode opposition. They would be right to do so—but they should pause. Deconstruction wields a powerful, wide-ranging, omnivorously aggressive vocabulary, and its adherents over the past two decades have appropriated so many different arguments,

conceptual moves, analyses, alternative logics, rethinkings, critiques, and overturnings as "deconstructions" that the word is now too overfed to be of much value. Similar remarks apply to "supplementarity." Undoubtedly, there is a sense in which Derrida's characterization of "writing" as a supplement to speech inside a logocentrized Western thought, as a "debased, lateralized, repressed, displaced theme, yet exercising a permanent and obsessive pressure from the place where it remains in check" (1976: 270), can be mapped onto the picture I draw of the metaCode's supplementary relation to the Code as this relation exists in the mind of the mathematical community. But a lot more than this kind of generic move is involved. Two points then. In the first place, the analysis here works through an inversion of Derrida's terms. What operates in mathematics is not logocentrism, not the privileging of speech over writing, of primary self-presence over a despised secondarity, but the reverse: a form of graphocentrism, the privileging of the formal writing of the Code over an eliminable, theoretically unnecessary—epiphenomenal—metaCode. Moreover, this privileging of Coded writing over metaCoded speech is itself internalized within the metaCode where it emerges as a refusal of ostension, a kind of avoidance, if not a phobia, of body signals: if formal writing is valued over speech, then speech, however undesirable, is seen as superior to the atavism of pre-speech, to gesture, to the ultimate negative epithet in the policing by mathematicians of their subject's boundaries: mere *handwaving*. In the second place, Derrida himself seems to conceive of mathematics—because its notations are ideogrammatic and not alphabetic—as a model and ideal candidate for the writing that he advocates, one free of logocentrism and untainted by the metaphysics of presence (1981b: 34–35). Against (or at least hardly for) this, the entire thrust of my essay is not only to locate a deeply metaphysical principle at work within mathematics' current conception of "number" but also to reinstate the body and the subject with their talk, noise, and physics of *presence* onto the mathematical scene.

29. Thought experiments are so intrinsic to the mathematical scene that they are likely to appear—unnoticed and consequently untheorized—simply as the terms in which one describes "mathematical thought" itself. Thus, in their list of "intellectual components of mathematical thinking," Davis and Hersh give as two principal items "ability and willingness to extract out of the real an abstract surrogate; correspondingly, the willingness to accept formal manipulations of the abstract surrogate as an adequate representation of the real," and "ability and desire to manipulate and play with symbols even in the absence of concrete referents, thus creating an imaginary world which transcends the concrete"—the means and wish, in other words, to cognize thought experiments (1988: 124–25).

30. Evidently, not all mathematical assertions yield so directly and simply to a construal in terms of predictions as the examples used here. Thus, to secure the point in any substantive way, one would need to formulate the appropriate type of general argument—one that would have to be based, say, on the recursive construction of mathematical propositions from the sort of simpler ones given here. This seems to me a technically onerous but not theoretically problematic task, and I have in the text assumed that it or something like it could be carried out.

31. "Intensional" here has the sense of being description-dependent. In the present context, the description is the work of the Person—the one able to recognize that two proofs of the same assertion are not inter-changeable and able to articulate this in the metaCode. Proofs can and do differ in terms of scope, linguistic means, intelligibility, complexity, elegance, intention, artificiality, redundancy, generalizability, obviousness, novelty, narrative structure, appropriateness, naturality, fruitfulness, perspicacity . . . —and these differences frequently play a leading and un-eliminable role in the ongoing construction of mathematics.

32. The impact of mathematics' instrumentality on its own development alluded to here (a topic we shall return to in Chapter 5) is evident enough. Less obvious, perhaps, is the vector from the publicly constituted meta-Code to the subCode of the Agent: indeed, the institution and legitima-tion of the Agent—what can be assumed about its capacities to repeat, to select, to survey, and so on—has received practically no attention from the mathematical community. A notable exception would be the debates in the first quarter of this century surrounding the Axiom of Choice: the issue being whether a simultaneous infinity of choices could or could not be credited to the Agent, and if it could, whether it made sense for that infin-ity to be nondenumerable. See Sierpinski (1958: 88–95) for an excellent summary of the issues under debate.

33. For further remarks about functors and category theory see note 41. A much fuller account (which makes no reference to categories and func-tors) of the way the model here delivers a unified, triadic critique of the guiding assumptions behind intuitionism's, Platonism's, and formalism's use of signs can be found in Rotman 1988.

34. The present account was conceived and formulated before I had read Kitcher 1984, where a model of arithmetic—on the face of it remark-ably similar to the one here—as an "idealizing theory" in which "arith-metic owes its truth, not to the actual operations of actual human agents, but to the ideal operations performed by ideal agents" (109) is put for-ward. In addition, Kitcher argues for an empirically based view of mathe-

matics grounded on the historical genesis of mathematical abstractions having much in common with the underlying assumptions here. Kitcher's essay, which is illuminating and thoughtfully argued, deserves extended comment in relation to the present enterprise. There are, however, two factors that might make a comparison between his and my account somewhat unproductive and artificial. First, Kitcher's purpose is conservative and nonrevisionary with respect to infinity: he accepts, with no apparent qualms or evidence of struggle, the endless supply of natural numbers, with the aim not of challenging but rejustifying the existing body of mathematics and the classical logic that supports it. Second, he works within a classical philosophical framework that separates a prior ontology of objects from a subsequent epistemology and takes as its central concept "knowledge" based on "truth" about mathematical existents assumed to be independent of the knowing subject. By contrast, the present account sees the mathematical world and all that constitutes mathematical "reality" as constructed by the very same linguistic/semiotic means used by subjects to "describe" it—thus making "knowing" a self-reflexive process difficult to recognize and articulate—and impossible to theorize—within the classical framework.

35. The topos of "nothing for nothing" which Gabor's remark and my appropriation of it exemplify, though ancient and of wide-ranging application—from "nothing will come of nothing" to "there's no such thing as a free lunch"—has a particular anti-objectivist edge here. What it denies is the existence of an unresourced outside, an exterior transcendental realm, field of "pure" activity where processes occur and things happen without cost, without reckoning, without all the reciprocations of price—whether in terms of money, obligation, energy, or whatever—that testify to the ineluctably messy immersion in empirical materiality of all realizable (actual or imagined) human transactions. In her book *Contingencies of Value* Smith calls the particular version of this objectivist fantasy that hovers over accounts of value and worth the "dream of an escape from economy" and marvelously charts its dynamics and its reliance on a "superlunary universe" which even then would still require an "archangel perhaps" (1988: 111) to somehow introduce difference into such a realm. My gloss on the final state of Maxwell's Demon was written before I was aware of the important work of Rolf Landauer (1986) and Charles Bennett (1987) on the physical limits of computation.

36. Indeed, the possibility exists of an appropriation by classical mathematics of the conception of number presented here. Might one not make classical sense of the limit denoted by $—representing it in terms of an

entropic, degenerative *model* of the classically, preexistent natural numbers? Thus, we could construe the dissipation in terms of the effect of a (classically described) function *f* that mapped the sequence of infinitistic natural numbers onto the sequence of realizable iterates, in such a way that $f(n)$ would be the *n*th iterate, and $f(x)$ would tend to 0 as *x* increased. But such a response, assuming it could be made coherent, would be beside the point: the dissipation here is an internal and self-sustaining effect of the realizability constraint and makes itself felt on *all* realizable numbers. And it is difficult to imagine how any purely external, classically interpretable mapping of 0, 1, 2, . . . onto 0, 1, 2, . . . $ could preserve this internal structure without generating a contradiction. Thus, for example, from a classical perspective starting one's counting process from 1 is no different from starting it from *n* for any number *n*. For realizable numbers this is patently not the case. And the question would be how any classical appropriation of $ could avoid a contradiction here.

37. The general question of physics' de jure *need for*, as opposed to its de facto *use of*, infinitary mathematical techniques is a large and important issue that remains unsatisfactorily explored. Certainly, considered just in terms of the measurements it makes and the data values it uses to theorize reality, physics has no need, in principle, ever to go beyond the small subset of rational numbers determined by the limits of accuracy—30 decimal places, say—it operates with. For an initial understanding of the consequences for physics of employing a "deinfinitized"—interpreted as "realizable"—arithmetic, see the discussion of measurement in Chapter 5 below.

38. Such an account would be part of a much larger examination of who or what speaks, and who or what is the intended hearer, of scientific writing. See Markus 1987 for the general nature of the sort of examination required, as well as some suggestions as to why it has so far not been forthcoming. Many of the considerations advanced by Markus, particularly in relation to the constitution of the ideal reader/writer of scientific texts, are directly applicable to mathematical texts.

39. See Kitcher (1984: 41–46) for a discussion of what caused Descartes and Hume to worry about the cognition of extended logical arguments. In Kitcher's view these worries are basically justified, and he interprets the points they raise as further evidence against what he calls aprioristic accounts of mathematical knowledge. A different kind of response, more in tune with the outlook presented here, would be to investigate the consequences for arithmetic that issue from refusing to cognize arbitrarily

long proofs. See, for example, Parikh 1971, where a cut-off point to the length of a permissible mathematical argument is the starting point. Unfortunately, however, Parikh works with an entirely classical notion of "number," making his results not assimilable, at least not in any straightforward way, to the case being presented here.

40. Maslow's rule is cited in McLuhan and McLuhan (1988: 179). A similar law of satiety was formulated—unsatisfactorily, according to present-day economic orthodoxy, which holds that the concept of "utility" is not directly quantifiable—as the law of diminishing marginal utility, within the economic theories of Hermann Gossen and again by William Jevons and Alfred Marshall at the end of the nineteenth century; see Bannock et al. (1972: 193–94).

41. The term "functor" as well as the descriptions "forgetful" and "limit" are taken from category theory. What is being suggested here is that if "entity" is interpreted as "category" and "Y models X" as the existence of a pair of linked functors

$$X \underset{q}{\overset{p}{\rightrightarrows}} Y$$

where p is a forgetful functor and q a limit or idealization functor, then, on general category-theoretic grounds, there will be an "inverse" functor p^* to p, known as the left-adjoint of p; and, under the appropriate assumptions, a right-adjoint functor $q^\wedge$ "inverse" to q, known as the diagonalization functor. This means that one can embed "Y models X" within a richer picture of relations:

$$X \underset{p^*,q^\wedge}{\overset{p,q}{\rightleftarrows}} Y$$

Applying these ideas to the situation under discussion, yields the following diagram:

$$\text{Person} \underset{a^*,c^\wedge}{\overset{a,c}{\rightleftarrows}} \text{Subject} \underset{b^*,d^\wedge}{\overset{b,d}{\rightleftarrows}} \text{Agent}$$

where a and b are forgetful functors and c and d are limit functors. The interpretation of a^* and b^* would be in terms of "free algebras" of metasigns generated by signs and "free algebras" of signifieds generated by signifiers, respectively; and the interpretation of $c^\wedge$ and $d^\wedge$ would be by way of the construction of "limit objects" governing the retrieval of metaCode from Code, and Code from subCode, respectively. See Mac-Lane 1971 for the definitions and properties of adjoints and diagonalization functors. One possible use of the foregoing category-theoretic formulation (the development of which would take us far outside the bounds of the present work) would be to provide a lower bound to the complexity of any proposed model of an "artificial intelligence" held capable of doing mathematics.

42. On this question of the interconnection of metaCode and the Agent's subCode, compare note 32 as well as the discussion in Chapter 5 concerning the two-way traffic between Person, Subject, and Agent.

43. Skepticism about the status of the standardly conceived real numbers as adequate, useful, "true" descriptions of reality is by no means unknown. Thus, see Eddington (1935: 103) for a (not too convincing) retraction of his earlier more penetrating doubts in relation to the micro-distances required by quantum theory. More recently, and independently of Eddington, the question is raised by Penrose (1989: 86–87) who finds it counterintuitive to imagine physical situations answering to arbitrarily small lengths and time intervals. As he points out (with a candor about the issue unusual among physicists), there is no justification for the prevailing confidence in the applicability of the real numbers far removed from the "comparatively limited range" of our experience. Indeed there is not; and we shall take this lack of justification seriously, on the level not just of the real numbers but of elementary arithmetic itself from which all conceptions of real numbers are constructed, when we come to set up non-Euclidean arithmetic below. In this we shall formulate a general program and arrive at conclusions remarkably similar to those adumbrated by Rashevskii 1973, which, like the present approach, starts from a radical questioning of the relation between counting and the so-called endlessness of the natural numbers.

44. The study of nonlinear behavior of dynamical systems has undergone a vast expansion in the last two decades, resulting in many different kinds of systems and patterns of equilibria. My comparison between the structure of realizable arithmetic and that of a dynamical system is, therefore, extremely generic. Notwithstanding this, the kinds of structures I have in mind and the bifurcations they exhibit are drawn from the discussion of the behavior of open systems in far-from-equilibrium regions by Prigogine and Stengers (1984).

45. One might compare the claim here of the—necessary—existence of such a numerical information horizon around any singularity with the—ad hoc—"principle of cosmic censorship" put forward by Penrose (1989:

215), according to which the hypothesized existence of an event horizon (of an associated black hole) prevents our ever encountering a "naked" singularity.

46. From the classical point of view the realizable continuum would appear to be full of "gaps" and the boundaries described in the text would be upper and lower bounds to a classically delimitable subset of the real numbers. Thus, consider the following concrete working expression, taken from the manual for a computer language, of the classical view in relation to numbers as handled by a computer: "The real number system is both infinitely long and infinitely dense. In addition to stretching from negative to positive infinity, between any two points on the number line there are an infinity of points [numbers]. . . . Finite machines that they are, computers cannot exactly represent this transcendent system of numbers; . . . there are gaps between adjacent representable values [which] get larger as you move further from zero in either direction on the number line. Eventually, you get so far away from zero that there are no representable points left. Producing numbers beyond the last representable value in either direction is called *overflow*" (Borland Corp. 1989). According to the present account such "gaps" are nothing of the kind; they correspond instead to inescapable facts of (realizable) arithmetic life—namely, there exist adjacent transrationals. The realizable continuum can thus be pictured by thinking of the above description as referring to the idealized Limit machine, where of course "infinitely long" has to be replaced by $-long, "infinitely dense" has to be rejected, "representable" means simply "exists," and "gaps" are replaced by "neighbors."

47. Max Planck is quoted in Barrow and Tipler (1988: 292) where derivations of the various Planck units of length, time, mass, and temperature are given, along with a discussion of the history of attempts to arrive at such "natural" units.

48. In other words, what is *not* desirable (except by Platonists perhaps), since it only writes into its account the very occlusion of human agency that forms its story, is the history of mathematics as the accumulation of "true" knowledge. One influential repudiation of this form of progressive rationalism is the approach of Imre Lakatos (1976). Compare also the arguments of Raymond Wilder (1981 and many of his earlier works), that mathematics should be seen as a culturally and linguistically produced system of thought. Somewhat differently, Philip Kitcher (1984) argues for the importance of historical causation to any empiricist account of mathematics' influence and knowledge claims. On a less subject-wide scale, there is a growing recognition by some historians of science of the technical

construction of reason, of the way the applications of mathematics and the elaboration of standards of "objectivity" which facilitate the acceptance and naturality of these applications, go hand in hand. See, for example, Porter 1990, where the phenomenon is discussed in relation to accountancy practices.

49. Autogenic feedback, the circular movement between internal theory and growth and external practice and technical applications, should be compared to the biological mechanism of *autopoeisis* articulated by Humberto Maturana to theorize the achievement of a successful "structural coupling" between an organism and the environmental medium in which it has its being; see Maturana and Varela 1980. See also Winograd and Flores 1986 for a use of autopoeisis in the service of a critique of the "rationalistic tradition."

50. It is perhaps inevitable that any principle of epistemic change predicated on the action of negation will suggest Hegel's dialectical move from thesis to synthesis via negation of the thesis, but this suggestion should be resisted. What I call epistemic foreclosure is a phenomenon of *cognito-semiotic* opening and novelty—the undoing and de-thinking of an *X* that was previously a finished and closed thought—not an effect of a supposedly transcendental logic working a passage through to a higher and necessary synthesis. (Moreover, though with a different emphasis from the text here, the principle can also be seen to operate within the emergence of zero: here the denied *X* is the assumption of a perfect and closed adequation, a matching of number signs and the world of palpable—but of course ideal—collections that preexist any numerical intervention; and with the rejection of this adequation comes the introduction of the one-who-counts [Rotman 1987]. And, for all its present-day "inevitability," the introduction of zero into Western culture can hardly be fitted under the rubric of an inexorable dialectical advance.) Nearer the mark one might want to interpret the negation—*denial* of *X*—at work here in terms of Freud's use of *Verneinung* as a way of taking cognizance of what is repressed. Thus, the introduction of the Corporeal Subject into mathematics represents the lifting of the repression of the body, a repression of mortality inseparable from the desire for endless counting. But this only goes so far: epistemic foreclosure has to be understood as a socio-cultural practice: it concerns discourse, codes, communication, and writing in relation to an historically shaped community. It cannot be formulated in terms of the Freudian dynamics of an individual psyche (even if this psyche is made into a panindividual object as in Lacan's understanding of the unconscious as being structured like a *language*, since there is still no history, culture,

or open-ended intersubjectivity in such an understanding). So that if one is to talk of repression, it has to be at the level of cultural practice—where it appears as ideology and hidden or naturalized metaphysics. Then what is involved concerns not so much the repression of individual deaths— a banal necessity without which one cannot count very far, let alone "endlessly"—but the death of the deity who can so count: a question of Nietzsche, then, rather than Freud.

51. For a typically graphic and illuminating reference to one such area of troublesome infinities—the difficulties of "arbitrarily close" positions that seem inseparable from the attempt to sum over all possible histories of a particle in quantum physics—see Feynman 1988: 129.

52. Psychoanalytically, nothing—especially an overarching desire for order—is innocuously itself. But I have not attempted to fold into the idea of mathematics-as-dream any psychoanalytical dimension, a theory of the dreaming—unconscious—subject, for example. No doubt this was a re- ductive and, for some perhaps, not the most interesting, decision, but it seemed to me, given the case about infinity I was pursuing, that any inclu- sion of Freud's conception of the unconscious and Lacan's Hegelian rewrit- ing of that conception would have needed such a degree of qualification (see note 50 above) as to have considerably muddied the water here with- out providing any compensating addition to the substance of my argu- ment. However, if one leaves the problematics of infinity aside, then the dream of order, regularity, repeatability, and control that mathematics offers—and with it the idea of a "pure," disembodied reason—clearly de- mands not only some sort of psychoanalytic account, but also one alive to the issue of gender that threads through any fantasy of self-empowerment through purity and control. See Walkerdine (1990: 183–216) for an inci- sive discussion of just this question in the context of the induction, edu- cational transmission, and social dissemination of mathematics.

53. The point here is that the total orderedness of the transiterates is not decided as an automatic consequence of the definition of realizable iterates, but has to be *posited*. In the absence of such a decision one has a variety of partially ordered, non-Euclidean models of arithmetic.

54. If one takes a totally ordered model of the transiterates, then there is a close resemblance between the structure of the integers as the ordered sum of iterates and transiterates given here and classically presented, count- able nonstandard models of arithmetic. Specifically, the iterates map onto the standard ω-sequence and the transiterates onto the η-product of $(\omega^* + \omega)$ sequences, where ω is the order-type of the natural numbers and η the order type of the rationals. To convert the resemblance into a

theorem, one would have to replace the (ω^* + ω) sequences by their realizable equivalents and likewise, since the realizable rationals are not dense, replace η with its realizable version. It is not clear, however, that any such *direct* comparison with classical metalogical results is plausible. Thus, to take a better known and very different example, the arithmetization of syntax required by Gödel's incompleteness theorem depends on the closure of the integers with respect to exponentiation: no such closure and so no corresponding arithmetization holds for the syntax of the realizable integers. In other words, some (by no means mechanical or obvious) work of relativization would have to be done before one could start trying to prove theorems (or provide counterexamples) about the structure of realizable arithmetic corresponding to (or contradicting) classical metamathematical results.

Amrine, Frederick, ed. 1989. *Literature and Science as Modes of Explana-tion.* Boston Studies in the Philosophy of Science, 115. Boston: Kluwer Academic Publishers.

Bannock, G., R. Baxter, and R. Rees. 1972. *Dictionary of Economics.* London: Penguin.

Barrow, John D., and Frank J. Tipler. 1988. *The Anthropic Cosmological Principle.* Oxford: Oxford University Press.

Bateson, Gregory. 1972. *Steps to an Ecology of the Mind.* New York: Ballantine Books.

Benardete, Jose A. 1964. *Infinity: An Essay in Metaphysics.* Oxford: Oxford University Press.

Bennett, Charles. 1987. "Demons, Engines and the Second Law." *Scientific American* 257 (5): 108–16.

Benveniste, Emile. 1971. *Problems in General Linguistics.* Trans. Mary E. Meek. Coral Gables, Fla.: University of Miami Press.

Berlin, Isaiah. 1976. *Vico and Herder.* London: The Hogarth Press.

Bernays, Paul. 1964. *On Platonism in Mathematics.* Philosophy of Mathe-matics. Englewood Cliffs, N.J.: Prentice-Hall.

Bishop, Everett. 1967. *Foundations of Constructive Analysis.* New York: McGraw-Hill.

Bloor, David. 1976. *Knowledge and Social Imagery.* London: Routledge.

Bloor, David, and Barry Barnes. 1982. "Relativism, Rationalism and the Sociology of Knowledge." In Martin Hollis and Steven Lukes, eds. 1982. *Rationality and Relativism.* Cambridge, Mass: Harvard University Press.

Borland Corp. 1989. *Turbo Basic Owner's Handbook.*

Bourdieu, Pierre. 1977. *Outline of a Theory of Practice.* Cambridge, Eng.: Cambridge University Press.

Boyer, Carl. 1968. *A History of Mathematics.* New York: John Wiley.

Bridges, Douglas S., and Fred Richman. 1987. *Varieties of Constructive Mathematics.* Cambridge, Eng.: Cambridge University Press.

Brown, James Robert. 1991. *The Laboratory of the Mind: Thought Experiments in the Natural Sciences*. London and New York: Routledge.

Buchler, Justus. 1940. *The Philosophy of Peirce: Selected Writings*. London: Routledge.

Culler, Jonathan. 1981. *The Pursuit of Signs*. London: Routledge.

Dantzig, Tobias. 1953. *Number: The Language of Science*. London: Macmillan.

Davis, Phillip J., and Reuben Hersh. 1983. *The Mathematical Experience*. London: Penguin.

_____. 1988. *Descartes' Dream*. London: Penguin.

Dedekind, Richard. 1963. *Essays on the Theory of Numbers*. New York: Dover Books.

Derrida, Jacques. 1976. *Of Grammatology*. Trans. Gayatri C. Spivak. Baltimore, Md.: The Johns Hopkins University Press.

_____. 1981a. "Economimesis." *Diacritics* 11: 3–25.

_____. 1981b. *Positions*. Chicago: University of Chicago Press.

_____. 1982. *Margins of Philosophy*. Trans. Alan Bass. Chicago: University of Chicago Press.

_____. 1989. *Edmund Husserl's 'Origin of Geometry': An Introduction*. Trans. John P. Leavey. Lincoln: University of Nebraska Press.

Drake, Stillman. 1957. *Discoveries and Opinions of Galileo*. New York: Doubleday.

Dummett, Michael. 1975. *What Is a Theory of Meaning?* vol. 1. Oxford: Oxford University Press.

Eddington, Arthur S. 1935. *New Pathways in Science*. Cambridge, Eng.: Cambridge University Press.

Ernest, Paul. 1991. *The Philosophy of Mathematics Education*. New York: The Palmer Press.

Feynman, Richard. 1988. *QED*. Princeton, N.J.: Princeton University Press.

Frege, Gottlob. 1967. "The Thought: A Logical Enquiry." In Peter F. Strawson, ed., *Philosophical Logic*. Oxford: Oxford University Press.

Gabor, Dennis. 1964. "Light and Information." *Progress in Optics* 1: 109–53.

Galileo Galilei. 1953. *Dialogue Concerning the Two Chief World Systems*. Trans. Stillman Drake. Berkeley: University of California Press.

Gardner, Martin. 1983. *The Whys of a Philosophical Scrivener*. New York: Wm. Morrow.

_____. 1989. *The New York Review of Books*, March 16, pp. 26–28.

Goodman, Nelson. 1978. *Ways of Worldmaking*. Indianapolis: Hackett.

Greenlee, Douglas. 1973. *Peirce's Concept of the Sign*. The Hague: Mouton.

Harris, Roy. 1986. *The Origin of Writing*. London: Duckworth.

Hughes, Patrick, and George Brecht. 1978. *Vicious Circles and Infinity*. London: Penguin Books.

Husserl, Edmund. 1989. *The Origin of Geometry*. In Jacques Derrida. 1989. *Edmund Husserl's 'Origin of Geometry': An Introduction*. Trans. John P. Leavey. Lincoln: University of Nebraska Press.

Isles, David. 1981. "On the Notion of Standard Non-isomorphic Natural Number Series." In Fred Richman, ed., *Constructive Mathematics*. New York: Springer-Verlag.

_____. Forthcoming. "What Evidence Is There That 2^{65536} Is a Natural Number?"

Johnson, Mark. 1987. *The Body in the Mind: The Bodily Basis of Meaning, Imagination, and Reason*. Chicago: University of Chicago Press.

Kant, Immanuel. 1986. *The Critique of Judgement*. Trans. George Meredith. Oxford: Oxford University Press.

Kitcher, Philip. 1984. *The Nature of Mathematical Knowledge*. New York: Oxford University Press.

Klenk, V. H. 1976. *Wittgenstein's Philosophy of Mathematics*. The Hague: Martinus Nijhoff.

Kuhn, Thomas S. 1977. *The Essential Tension*. Chicago: University of Chicago Press.

Lakatos, Imre. 1976. *Proofs and Refutations*. Cambridge, Eng.: Cambridge University Press.

Landauer, Rolf. 1986. "Computation and Physics: Wheeler's Meaning Circuit?" *Foundations of Physics* 16(6): 551–64.

Lorenzen, Paul. 1955. *Einführung In die operative Logik und Mathematlk*, vol. 1. Berlin: Springer-Verlag.

MacLane, Saunders. 1971. *Categories for the Working Mathematician*. New York: Springer-Verlag.

McLuhan, Marshall, and Eric McLuhan. 1988. *Laws of Media*. Toronto: University of Toronto Press.

Markus, Gyorgy. 1987. "Why Is There No Hermeneutics of Natural Sciences? Some Preliminary Theses." *Science in Context* 1: 5–51.

Maturana, Humberto R. and Francesco Varela. 1980. *Autopoiesis and Cognition*. Dordrecht: Reidel.

Moore, A. W. 1990. *The Infinite*. London: Routledge.

Nelson, Edward. 1986. *Predicative Arithmetic*. Mathematical Notes 32. Princeton, N.J.: Princeton University Press.

Nietzsche, Friedrich. 1956. *The Genealogy of Morals*. Trans. Francis Golffing. New York: Anchor Doubleday.

Parikh, Rohit. 1971. "Existence and Feasibility in Arithmetic." *Journal of Symbolic Logic* 36 (3): 494–508.

Parsons, Charles. 1983. *Mathematics in Philosophy*. Ithaca, N.Y.: Cornell University Press.

Pascal, Blaise. 1950. *Pensées*. Trans. H. F. Stewart. New York: Pantheon Books.

Pavis, Patrice. 1982. *Languages of the Stage*. New York: Performing Arts Journal Publications.

Pearson, Karl. 1937. *The Grammar of Science*. London: Dent.

Peirce, Charles S. 1931–58. *Collected Papers*, vols. 1–8. Ed. Charles Hartshorne and Paul Weiss. Cambridge, Mass.: Harvard University Press.

Penrose, Roger. 1982. Letter to *The Times Literary Supplement*, 11 June.

_____. 1989. *The Emperor's New Mind*. London: Penguin Books.

Poincaré, Henri. 1913. *The Foundations of Science*. New York: Science Press.

Pool, Ithiel de Sola. 1990. *Technologies Without Boundaries: On Telecommunications in a Global Age*. Cambridge, Mass.: Harvard University Press.

Porter, Theodore. 1990. "Quantification and the Accounting Ideal in Science." Paper delivered at the History of Science Society, Seattle, Oct. 19.

Prigogine, Ilya, and Isabelle Stengers. 1984. *Order out of Chaos*. London: Collins.

Putnam, Hilary. 1983. *Realism and Reason*. Vol. 3 of *Philosophical Papers*. Cambridge, Eng.: Cambridge University Press.

Rashevskii, P. K. 1973. "On the Dogma of the Natural Numbers." *Russian Mathematical Surveys* 28 (4): 143–48.

Restivo, Sal. 1983. *The Social Relations of Physics, Mysticism, and Materialism*. Boston: Reidel.

_____. 1990. "The Social Roots of Pure Mathematics." In S. Cozzens and T. Gieryn, eds. *Theories of Science in Society*. Bloomington: Indiana University Press.

Rorty, Richard. 1979. *Philosophy and the Mirror of Nature*. Princeton, N.J.: Princeton University Press.

Rothman, Tony. 1989. *Science à la Mode: Physical Fashions and Fictions*. Princeton, N.J.: Princeton University Press.

Rotman, Brian. 1987. *Signifying Nothing: The Semiotics of Zero*. London: Macmillan. Rpt. 1993. Stanford, Calif.: Stanford University Press.

_____. 1988. "Towards a Semiotics of Mathematics." *Semiotica* 72: 1–35.

Saussure, Ferdinand de. 1966. *Course in General Linguistics*. Trans. Wade Baskin. New York: McGraw-Hill.

Shapin, Steve, and Simon Schaffer. 1985. *Leviathan and the Air-Pump*. Princeton, N.J.: Princeton University Press.

Sierpinski, Waclaw. 1958. *Cardinal and Ordinal Numbers*. Warsaw: Panstwowe Wydawnictwo Naukowe.

Silverman, Kaja. 1981. *The Subject of Semiotics*. Oxford: Oxford University Press.

Smith, Barbara Herrnstein. 1988. *Contingencies of Value*. Cambridge, Mass.: Harvard University Press.

Sorensen, Roy A. 1992. *Thought Experiments*. Oxford: Oxford University Press.

Steiner, Mark. 1975. *Mathematical Knowledge*. Ithaca, N.Y.: Cornell University Press.

Stolzenberg, Gabriel. 1970. Review of Everett Bishop's *Foundations of Constructive Analysis*. In *Bull. Amer. Math. Soc.* 76: 301–23.

———. 1978. "Can an Inquiry into the Foundations of Mathematics Tell Us Anything Interesting About Mind?" In George Miller, ed. *Psychology and Biology of Language and Thought*. New York: Academic Press.

Tiles, Mary. 1989. *The Philosophy of Set Theory*. Oxford: Basil Blackwell.

Tragesser, Robert S. 1984. *Husserl and Realism in Logic and Mathematics*. Cambridge, Eng.: Cambridge University Press.

Van Bendegem, J. P. 1986. *Finite, Empirical Mathematics: Outline of a Model*. Seminarie voor Logica en Wijsbegeerte van de Wetenschappen, 17. Rijksuniversiteit, Gent, Belgium.

Van Dantzig, D. 1959. "Is $10^{\text{ten billion}}$ a Finite Number?" *Dialectica* 9: 273–77.

Walkerdine, Valerie. 1990. *The Mastery of Reason*. New York: Routledge.

Weyl, Hermann. 1949. *Philosophy of Mathematics and Natural Science*. Princeton, N.J.: Princeton University Press.

Wigner, Eugene P. 1960. "The Unreasonable Effectiveness of Mathematics in the Natural Sciences." *Communications on Pure and Applied Mathematics* 13: 1–14.

Wilden, Anthony. 1972. *System and Structure: Essays in Communication and Exchange*. London: Tavistock Publications.

———. 1987. *The Rules Are No Game: The Strategy of Communication*. London: Routledge.

Wilder, Raymond L. 1981. *Mathematics as a Cultural System*. Oxford: Pergamon Press.

Winnicott, Donald W. 1971. *Playing and Reality*. London: Penguin Books.

Winograd, Terry, and Fernando Flores. 1986. *Understanding Computers and Cognition: A New Foundation for Design*. Norwood, N.J.: Ablex Publishing Corp.

Wittgenstein, Ludwig. 1961. *Tractatus Logico-Philosophicus*. Trans. David Pears and Malcolm McGuiness. London: Routledge.

———. 1978. *Remarks on the Foundations of Mathematics*. Oxford: Basil Blackwell.

Wright, Crispin. 1980. *Wittgenstein on the Foundations of Mathematics.*
  London: Duckworth.

Yessenin-Volpin. 1970. "The Ultra-intuitionistic Criticism and Anti-traditional
  Program for Foundations of Mathematics." In A. Kino, ed. *Intuitionism
  and Proof Theory.* Amsterdam: North-Holland.

Library of Congress
Cataloging-in-Publication Data
Rotman, B. (Brian)

Ad infinitum—the ghost in
Turing's machine:
taking God out of
mathematics and putting
the body back in:
an essay in corporeal
semiotics / by Brian Rotman.
    p.   cm.
Includes bibliographical
references and index.
ISBN 0-8047-2127-0 (cloth :
alk. paper) : —
ISBN 0-8047-2128-9 (pbk. :)
1. Infinite.
2. Mathematics—Philosophy.
I. Title.
QA9.R775   1993
110—dc20
92-26420
CIP

This book is printed
on acid-free paper.

This book was typeset
using Quark XPress
by G&S Typesetters
in Adobe Stone Sans,
designed by Sumner Stone.
It was designed by
Copenhaver Cumpston.
Chapter opening art:
details from "Rupture" (1955)
by Remedios Varo.

CPSIA information can be obtained
at www.ICGtesting.com
Printed in the USA
JSHW031551160720
6747JS00002B/100